I0056308

ATOMIC AND MOLECULAR PHYSICS

LECTURE NOTES ON

ATOMIC AND MOLECULAR PHYSICS

Şakir Erkoç

Department of Physics
Middle East Technical University
06531 Ankara, Turkey

Turgay Uzer

Georgia Institute of Technology
Atlanta, Georgia 30332–0430, USA

World Scientific
Singapore • New Jersey • London • Hong Kong

Published by

World Scientific Publishing Co. Pte. Ltd.

5 Toh Tuck Link, Singapore 596224

USA office: 27 Warren Street, Suite 401-402, Hackensack, NJ 07601

UK office: 57 Shelton Street, Covent Garden, London WC2H 9HE

British Library Cataloguing-in-Publication Data
A catalogue record for this book is available from the British Library.

LECTURE NOTES ON ATOMIC AND MOLECULAR PHYSICS

Copyright © 1996 by World Scientific Publishing Co. Pte. Ltd.

All rights reserved. This book, or parts thereof, may not be reproduced in any form or by any means, electronic or mechanical, including photocopying, recording or any information storage and retrieval system now known or to be invented, without written permission from the publisher.

For photocopying of material in this volume, please pay a copying fee through the Copyright Clearance Center, Inc., 222 Rosewood Drive, Danvers, MA 01923, USA. In this case permission to photocopy is not required from the publisher.

ISBN-13 78-981-02-2811-8
ISBN-10 981-02-2811-2

Preface

The aim of this book is to present a unified account of the physics of atoms and molecules from a modern viewpoint. It is based on courses given by the authors at Middle East Technical University, Ankara and Georgia Institute of Technology, Atlanta. It is suitable for study at third and fourth year level of an undergraduate course.

It is our intent that students should be able to read this volume and understand its contents without the need to supplement it by referring to more detailed discussions. The whole subject covered in this volume is expected to be finished in one semester.

We hope that the worked examples at the end of each chapter will clarify some details in the chapters.

Ş. Erkoç
T. Uzer
May 1996

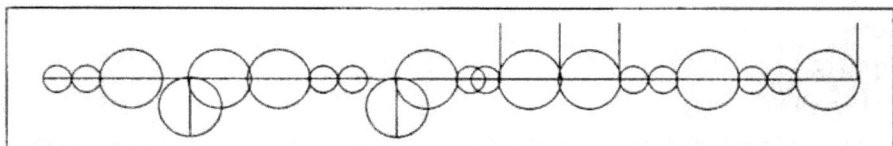

Contents

Chapter 1

Atomic Models

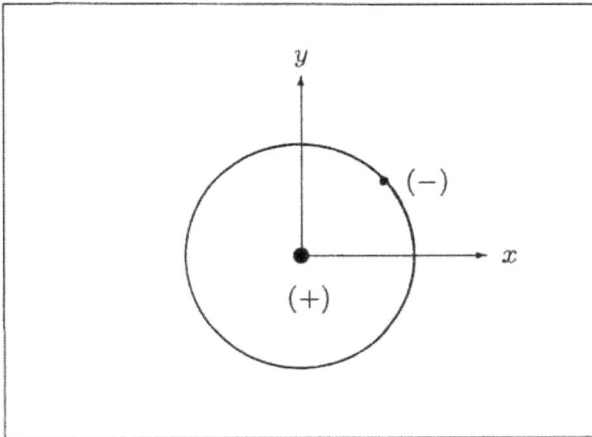

y

$(-)$

x

$(+)$

1.1   Rutherford's planetary model

Rutherford suggested that an atom is essentially a miniature solar system, with the electron moving in a plane along circular paths around the nucleus in analogy with the way that the planets orbit the sun.

For a stable orbit, the centripetal force between the two bodies must be supplied by the electrostatic force in the atom and gravitational force in the solar system.

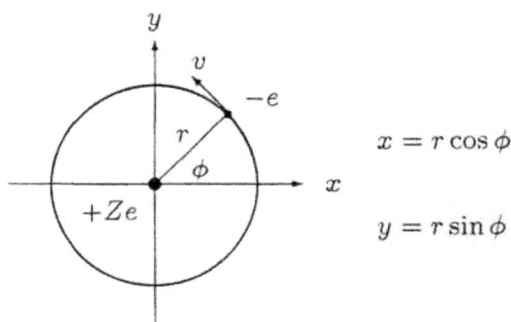

Fig. 1.1. Planetary model of hydrogen atom.

The electrostatic attraction between a nucleus of charge $+Ze$ and an electron of charge $-e$ gives rise to a potential energy of interaction equal to

$$V(r) = -\frac{Ze^2}{r} \tag{1.1}$$

Corresponding to this potential energy, there is a radial force F_r pulling the electron toward the nucleus, given by

$$F_r = -\frac{dV}{dr} = -\frac{d}{dr}\left(-\frac{Ze^2}{r}\right) = -\frac{Ze^2}{r^2} \tag{1.2}$$

If the electron is moving in a circular orbit with linear velocity v, the centripetal force is

$$F_c = \frac{mv^2}{r} \tag{1.3}$$

Using the condition for stability for a fixed nucleus with infinite mass, one can write

$$\frac{mv^2}{r} = \frac{Ze^2}{r^2} \tag{1.4}$$

The total energy E of the system is the sum of the kinetic energy T and the potential energy V of interaction; that is

$$E = T + V = \frac{1}{2}mv^2 - \frac{Ze^2}{r} \tag{1.5}$$

Solving Eq.(1.4) for r or for mv^2 and substituting into Eq.(1.5), we find

$$E = -\frac{1}{2}mv^2 = -\frac{Ze^2}{2r} \tag{1.6}$$

It is seen that the energy of the electron in the Rutherford atom is negative. This result from our choice of the zero of energy as the energy corresponding to the electron being infinitely far from the nucleus.

Eq.(1.6) is a special case of the **virial theorem**; it states that: **for a system of bound particles with gravitational or electrostatic forces acting between them, the total energy is equal to half of the average potential energy** or, equivalently, **to the negative of the average kinetic energy**.

In the Rutherford model V and T are constant for a given orbit, so that

$$E = \frac{1}{2}V = -T \tag{1.7}$$

In addition to the energy, the angular momentum also plays an important role in the theory of atomic structure. If ω is the angular velocity of the electron in its orbit about the nucleus, the angular momentum is

$$p_\phi = mvr = mr^2\omega \qquad (v = r\omega) \tag{1.8}$$

The kinetic energy can be rewritten as

$$T = \frac{1}{2}mv^2 = \frac{p_\phi^2}{2mr^2} \tag{1.9}$$

Combining Eqs.(1.6) and (1.9), one obtaines

$$r = \frac{p_\phi^2}{mZe^2} \tag{1.10}$$

This expression of r, the radius of the orbit, allows us to write the energy in terms of p_ϕ. From Eqs.(1.6) and (1.10), we have

$$E = -\frac{Ze^2}{2r} = -\frac{mZ^2e^4}{2p_\phi^2} \tag{1.11}$$

Although the planetary model is very simple, it contains a fundamental inconsistency which has so far been ignored. Since the electron is moving in a circular orbit, there is a constant acceleration corresponding to the continuous change in the direction of its velocity.

According to classical electrodynamics theory the accelerated electron must emit electromagnetic radiation of frequency ν equal to its circulation frequency

$$\nu = \frac{\omega}{2\pi} \tag{1.12}$$

and of an intensity proportional to the square of the acceleration, $a^2 = \omega^4 r^2$;

$$I = \left(\frac{2e^2}{3c^3}\right)a^2 = \left(\frac{2e^2}{3c^3}\right)\omega^4 r^2 \tag{1.13}$$

Since the atoms are continuously radiating in Rutherford's model, they are continuously losing energy. Thus, their energy becomes continuously more negative, as a result the radius continuously decreases until finally the electrons spiral into the nucleus. Furthermore, since such a spiraling, radiating electron has a continuously decreasing radius, its circulation frequency must also change continuously. From Eqs.(1.8) and (1.10) one can write the classical frequency

$$\omega = \left(\frac{Ze^2}{mr^3}\right)^{\frac{1}{2}} \quad or \quad \omega = \frac{mZ^2e^4}{p_\phi^3} \tag{1.14}$$

Thus, as r decreases from its initial value to zero, the frequency of the emitted radiation, which is equal to $\omega/2\pi$, varies continuously between the initial frequency and infinity. This conclusion is in disagreement with experimentally observed emission spectra.

At the end of the 19th Century, scientists had observed the hydrogen spectrum emitted by stars and found it to consist of very sharp, discrete frequencies. It is clear that the classical planetary model cannot be correct. Obviously a different approach is needed.

1.2 The Rydberg formula

In discussing atomic spectra, we take the hydrogen atom as an example because there is very detailed information about the hydrogen spectrum, on the other hand it is the simplest atom.

Balmer, in 1885, noted that the spacings of a series of lines could be expressed in simple algebraic form. If each line in the series is assigned an integer n, the frequency ν of the line is given quantitatively by the formula

$$\nu = const. \times \left(\frac{1}{2^2} - \frac{1}{n^2} \right) \quad ; \quad n = 3, 4, 5, ... \qquad (1.15)$$

The spacing between the lines becomes smaller with increasing n.

In practical spectroscopy, measurements are often made in terms of the wavelength λ, rather than the frequency. The relation between ν and λ is

$$\nu = \frac{c}{\lambda} \qquad (1.16)$$

where c is the speed of light. It is usual practice to express the measurements in terms of the reciprocal of the wavelength. This quantity, $\tilde{\nu}$, is called the wave number

$$\tilde{\nu} = \frac{1}{\lambda} = \frac{\nu}{c} \qquad (1.17)$$

The wave number is the number of waves per unit length; it is usually expressed in units of cm^{-1}. In terms of wave number, the Balmer formula is

$$\tilde{\nu} = 1.09 \times 10^5 \left(\frac{1}{2^2} - \frac{1}{n^2} \right) \quad cm^{-1} \qquad (1.18)$$

where the constant was estimated by Balmer from his observations of a particular series of lines.

The hydrogen spectrum containes many lines other than the Balmer series. Rydberg carried Balmer's idea further and made more accurate and extensive measurements. Rydberg found that all of the lines in the hydrogen spectrum could be obtained from a single general formula of the form

$$\tilde{\nu} = R \left(\frac{1}{2^2} - \frac{1}{n^2} \right) \tag{1.19}$$

where R is a constant (now called the Rydberg constant) and n_1 and n_2 are both integers, with $n_1 = 1, 2, 3, ...$ and $n_2 = n_1 + 1, n_1 + 2, n_1 + 3, ...$ Each n_1 value corresponds to a separate series, while the n_2 values are associated with individual lines of the series. Thus, $n_1 = 1, n_2 = 2, 3, 4, ...$ is a series which was originally discovered by Lyman and is called the Lyman series; $n_1 = 2, n_2 = 3, 4, 5, ...$ is the Balmer series; $n_1 = 3, n_2 = 4, 5, 6, ...$ corresponds to a series discovered by Paschen, and so on. For all of these series, the same constant R applies. For one–electron atoms other than hydrogen (for example, the ions $He^+, Li^{2+}, ...$), the spectrum is given by a formula having the same form as Eq.(1.19), but with a different constant in place of R. $R_{exp.} = 1.0967758 \times 10^5 cm^{-1}$ for hydrogen atom.

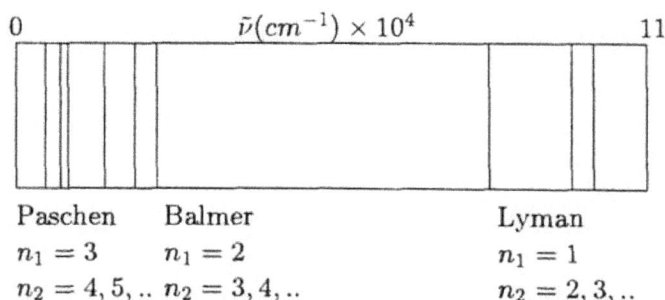

| 0 | $\tilde{\nu}(cm^{-1}) \times 10^4$ | 11 |

Paschen	Balmer	Lyman
$n_1 = 3$	$n_1 = 2$	$n_1 = 1$
$n_2 = 4, 5, ..$	$n_2 = 3, 4, ..$	$n_2 = 2, 3, ..$

Fig. 1.2. Lines in the hydrogen spectrum.

1.3   The Bohr theory of the atom

Rutherford's model predicts very short atomic lifetimes and continuous emission spectra which contradict experimental observations. To overcome these two problems and to provide a quantitative prediction of line spectra, N. Bohr proposed a new theory of atomic structure in 1913. He began with the planetary model, but introduced two general assumptions. The first assumption is that **an atom can exist for a long time without radiating in certain states with discrete energies**; these states are called stationary states. Secondly, Bohr assumed that **under certain conditions transitions between these stationary states do occur** and that these transitions are accompanied by the emission or absorption of radiation. These two assumptions state that the classical deductions from the planetary model do not apply to atoms.

Bohr's first assumption allows us to number the possible energies in increasing order starting with the one of lowest (most negative) value: $E_1, E_2, E_3, ...$ and so on. According to Bohr's second assumption, if an atom in one of the particular stationary states, say E_1, is placed in a beam of light, it may absorb enough energy from the radiation to make the transition to a stationary state of higher energy, say E_2. Conversely, an atom in an excited state, say E_2, may emit enough energy in the form of radiation to make a transition to a stationary state of lower energy, E_1. Planck had already used such a relationship to explain the observed variation with wavelenght of the intensity of radiation in equilibrium inside a hollow cavity (blackbody radiation). He had found that the energy ΔE absorbed or emitted is directly proportional to the frequency of the radiation, that is

$$\Delta E = h\nu \tag{1.20}$$

where the constant of proportionality h (Planck's constant) has the value $6.6256 \times 10^{-27} erg.s$. Planck's relation suggests that radiation is quantized; that is, it comes in energy packages $h\nu$, called quanta (one quantum) of energy. Bohr assumed that Planck's result applied to atoms with ΔE equal to the energy between two states; that is, the frequency of radiation emitted or absorbed by an atom as a result of a transition between the states E_{n_1} and E_{n_2} is determined by the frequency rule

(Bohr frequency rule)

$$\nu = \frac{E_{n_1} - E_{n_2}}{h} \quad or \quad \tilde{\nu} = \frac{E_{n_1} - E_{n_2}}{hc} \tag{1.21}$$

Comparison of Eqs.(1.19) and (1.21) suggests that the energies of the n_1 th and n_2 th stationary states are

$$E_{n_1} = -\frac{A}{n_1^2} \quad , \quad E_{n_2} = -\frac{A}{n_2^2} \tag{1.22}$$

where A is a constant equal to Rhc.

If the stationary–state energies are proportional to $1/n^2$ and planetary model is applicable, then the stationary–state radii and angular momenta must be proportional to n^2 and n, respectively. Letting a_0 and b_0 be the proportionality constants, we have

$$r = n^2 a_0 \quad , \quad p_\phi = n b_0 \tag{1.23}$$

The energy, radius, and angular momentum of a hydrogen atom are seen by the Bohr hypothesis to be restricted to a set of discrete values determined by the integer n. Therefore, we say that E, r, and p_ϕ are quantized, and refer to n as the quantum number. The smallest orbit ($n = 1$) with energy $E_1 (= -Rhc)$ has a radius a_0, the next orbit ($n = 2$) with energy $E_2 (= -Rhc/4)$ has a radius $4a_0$, and so on.

If a determination of any of the proportionality constants of Eqs.(1.22) or (1.23) were made independently of the experimental spectral data, the Rydberg constant R could be calculated theoretically. Bohr performed this very important step by introducing the correspondence principle, which was stated as in the limit of large quantum numbers, when the orbits become macroscopic, classical behavior must be approached; that is, the frequency of radiation for transitions between neighboring states, Eq.(1.21), should coincide with the classical frequency, Eqs.(1.12) and (1.14). Substituting for p_ϕ in Eq.(1.11) by Eq.(1.23) one obtains

$$E_n = -\frac{mZ^2 e^4}{2b_0^2 n^2} \tag{1.24}$$

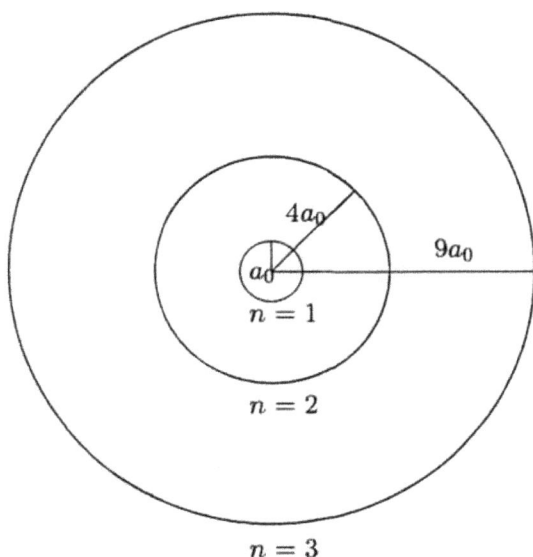

Fig. 1.3. Orbits of the electron in Bohr atom.

Introducing Eq.(1.24) into the Bohr frequency rule, Eq.(1.21), and setting $n_2 = n + 1$, and $n_1 = n$ we obtain

$$\nu_{n,n+1} = \frac{mZ^2e^4}{2b_0^2h} \left[\frac{1}{n^2} - \frac{1}{(n+1)^2} \right] = \frac{mZ^2e^4}{2b_0^2h} \left[\frac{2n+1}{n^2(n+1)^2} \right] \qquad (1.25)$$

According to the correspondence principle, as n becomes large, the right hand side of Eq.(1.25) must become equal to the classical frequency. For very large n,

$$\frac{2n+1}{n^2(n+1)^2} \overset{(n\to\infty)}{\longrightarrow} \frac{2}{n^3} . \qquad (1.26)$$

Thus, we can write

$$\nu_{n,n+1} \cong \frac{mZ^2e^4}{2b_0^2h} \left(\frac{2}{n^3} \right) \qquad (n \to \infty) \qquad (1.27)$$

If we now use Eq.(1.23) to reintroduce the angular momentum p_ϕ,

$$\nu_{n,n+1} \overset{(n\to\infty)}{\longrightarrow} \frac{mZ^2e^4b_0}{p_\phi^3h} \qquad (1.28)$$

Equating this limiting expression to the classical frequency given by Eqs.(1.12) and (1.14), we obtain

$$\frac{mZ^2e^4b_0}{p_\phi^3 h} = \nu_{class.} = \frac{mZ^2e^4}{2\pi p_\phi^3} \tag{1.29}$$

From this equation the constant b_0 must be set equal to $h/2\pi = \hbar$. Thus, from Eq.(1.23),

$$p_\phi = n\hbar \quad , \quad n = 1, 2, 3, ... \tag{1.30}$$

that is, the angular momentum is restricted to integer multiples of $\hbar$, for this reason $\hbar$ is sometimes called the **quantum of angular momentum**.

The radius for the n th stationary state orbit can be calculated by substituting Eq.(1.29) into Eq.(1.10); the result is

$$r = \frac{n^2\hbar_2}{mZe^2} \tag{1.31}$$

Comparison of Eqs.(1.31) and (1.23) shows that for the hydrogen atom $(Z = 1)$, the smallest radius corresponding to the stationary state of lowest energy $(n = 1)$ is

$$a_0 = \frac{\hbar^2}{me^2}. \tag{1.32}$$

This constant is called the **Bohr radius**, $a_0 = 0.529167 \times 10^{-8}$ *cm*. Substituting the Eqs.(1.31) and (1.32) into Eq.(1.11) one obtaines

$$E_n = -\frac{mZ^2e^4}{2\hbar^2}\left(\frac{1}{n^2}\right) = -\frac{Z^2e^2}{2a_0}\left(\frac{1}{n^2}\right) \quad , \quad n = 1, 2, 3, ... \tag{1.33}$$

Thus the quantized quantities are energy $E_n = -Z^2e^2/2a_0n^2$, momentum $p_n = n\hbar$, and radius $r_n = a_0n^2$.

The stationary state of lowest (most negative) energy, which is often called the ground state, is the state for which $n = 1$. For hydrogen $Z = 1$, the ground state energy is

$$E_1 = -2.18 \times 10^{-11} \; erg = -13.6 \; eV = -0.5 \; a.u. \tag{1.34}$$

This means that 13.6 eV of energy is required to take an electron from its most stable stationary state in the hydrogen atom ($n = 1$) and excite it to a state with $n \to \infty$. Since this final state is one for which $r(= a_0 n^2)$ is infinity, the potential energy of interaction between the electron and the nucleus is zero. The electron is now free of the nucleus and the atom is said to have been ionized. The energy supplied for the ionization process is called the ionization potential (IP). Ionization potential is a positive quantity, thus the negative of the ground–state energy is equal to the ionization potential for hydrogenic systems, namely $IP = -E_1$.

We can use the energy–level formula for the calculation of the Rydberg constant. Combining Eq.(1.33) with the frequency rule, Eq.(1.21), we have

$$\tilde{\nu} = \frac{1}{hc} \left(\frac{mZ^2 e^4}{2\hbar^2} \right) \left(\frac{1}{n_1^2} - \frac{1}{n_2^2} \right) \tag{1.35}$$

Setting $Z = 1$ for hydrogen

$$R = \frac{1}{hc} \left(\frac{me^4}{2\hbar^2} \right) = 1.097 \times 10^5 \ cm^{-1} \quad , \quad or \quad R = \frac{me^4}{4\pi c\hbar^3} \tag{1.36}$$

and

$$\tilde{\nu} = 1.097 \times 10^5 \left(\frac{1}{n_1^2} - \frac{1}{n_2^2} \right) \ cm^{-1} \tag{1.37}$$

in agreement with the experimental value.

It is now clear that the lines of the emission spectrum in each of the series correspond to transitions from various excited levels n_2 to a final state specified by the quantum number n_1. Thus, the Lyman series ($n_1 = 1$) consists of all transitions to the ground state ($E_2 \to E_1$, $E_3 \to E_1$, $E_4 \to E_1$, etc.), the Balmer series ($n_1 = 2$) consists of all transitions to the first excited state ($E_3 \to E_2$, $E_4 \to E_2$, $E_5 \to E_2$, etc.), the Paschen series ($n_1 = 3$) consists of all transitions to the second excited state ($E_4 \to E_3$, $E_5 \to E_3$, $E_6 \to E_3$, etc.), and so on.

The Bohr theory so far appears very satisfactory, since it is able to explain the sizes of atoms and the hydrogen spectrum quantitatively. Furthermore, it is found that Eq.(1.35) applies as well to the other one–electron atoms, if the value of Z corresponding to the nuclear charge is used; that is, $Z = 2$ for He^+, $Z = 3$ for Li^{2+}, and so on.

Returning to the energy equation, Eq.(1.33), we see that the energy levels of a one–electron atom are given by $E_n = -I_p/n^2$, where $I_p = Z^2 e^2/2a_0 = 13.6Z^2$ eV. If the atom absorbs an energy greater than I_p, the energy of the electron becomes positive and the electron is ejected from the atom. The quantity I_p is known as the **ionization potential**.

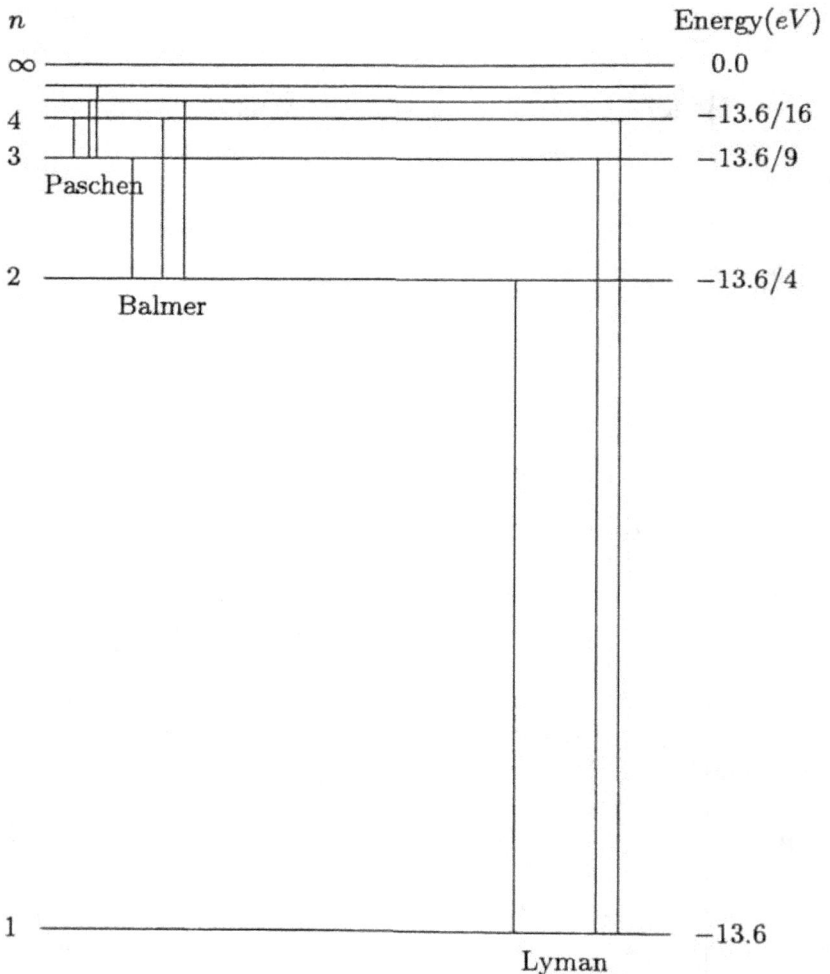

Fig. 1.4. Energy levels and transitions of the hydrogen atom.

1.4 The Frank–Hertz experiment

The Bohr model predicts that the energy levels of atoms are quantized. This fact was confirmed by an experiment originally devised by J. Frank and G. Hertz in 1914. A schematic diagram of the experimental setup is as shown in the figure.

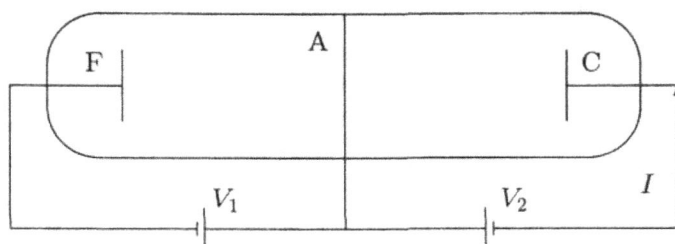

Fig. 1.5. Schematic diagram of Frank–Hertz experiment.

A filament F in a vacuum tube is heated so that it emits electrons, which are attracted to and pass through a grid A, which has a positive potential V_1 with respect to F. The kinetic energy of electrons may be determined from

$$\frac{1}{2}mv^2 = eV_1 \tag{1.38}$$

The electrons passing through the grid are collected by a plate C and cause a current I to flow in the circuit. The plate C has a small negative voltage V_2 with respect to A, $|V_2| \ll V_1$. The potential V_2 reduces the kinetic energy of the electrons but does not stop them.

The tube is now filled with mercury vapour. The electrons collide with the atoms of mercury, and if the collisions are elastic, so that there is no transfer of energy from the electrons to the internal structure of the atoms, the current I will be unaffected by the introduction of the gas. If an electron makes an inelastic collision with a mercury atom in which it loses an energy E, exciting the mercury atom to a level of greater internal energy, then its final kinetic energy will be

$$\frac{1}{2}mv_1^2 = eV_1 - E. \tag{1.39}$$

The experiment is carried out by gradually increasing V_1 from zero and measuring the current I as a function of V_1. The result obtained is as shown in Fig. 1.6.

The current I is seen to fall sharply at a potential V_R, which is known as the resonance potential, $V_R = 4.9\ eV$ for mercury. The result can be interpreted by supposing that for $eV_1 < 4.9\ eV$ the atom cannot absorb the energy of the electrons and the collisions are elastic, while exactly $4.9\ eV$ above its ground state, mercury atoms possess another discrete energy level. When eV_1 reaches this value, a large number of the colliding electrons excite atoms to this level, losing their energy in the process, and reducing the current I sharply.

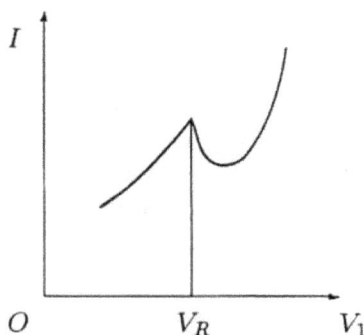

Fig. 1.6. Current vs voltage in Frank–Hertz experiment

If the voltage V_1 is increased the current again increases, and further sharp falls are seen. Some of these are due to electrons having sufficient energy to excite two or more atoms to the $4.9\ eV$ level, but others are due to the excitation of higher discrete levels.

This experiment provides confirmation of the discrete nature of bound state energy levels. It can be demonstrated that when sufficient energy is available to ionise an atom, the energy of the ejected electron can take any positive value, so we can say that the energy level spectrum of an atom consists of two parts: discrete negative energies corresponding to bound states and a continuum of positive energies corresponding to unbound states.

1.5 Correction for finite nuclear mass

Although the Bohr theory is successful for one–electron atoms, a small refinement is necessary. The need for the refinement becomes apparent experimentally when it is discovered that the frequency of the lines in the helium ion spectrum are not given exactly by multiplying the hydrogen spectrum by 4, as Eq.(1.35) suggests. The reason of the discrepancy is the assumption that the nucleus is stationary and that only the motion of the electron about the nucleus contributes to the energy of the system. This would be true only if the nucleus had infinite mass. In actual fact, the nucleus of the hydrogen atom has about 2000 times the electron mass, while for helium the ratio is about 6000 or 8000, depending upon the isotope.

Since both the nucleus and the electron have finite mass, they move about their common center of mass (C.M.). To obtain the Rydberg constant for a system with a nucleus of finite mass, we must rederive relations similar to Eqs.(1.4), (1.8), and (1.9), from which the Bohr formula was shown to follow.

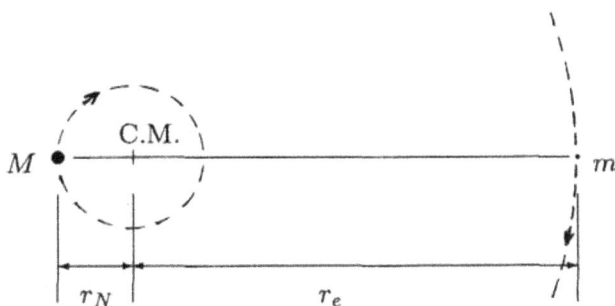

Fig. 1.7. Relative motions of p and e^- w.t. to C.M.

Consider an electron of mass m and a nucleus of mass M rotating about their C.M. The distances are related by

$$M r_N = m r_e \quad and \quad r = r_N + r_e \tag{1.40}$$

or

$$r_e = \left(\frac{M}{M+m}\right) \quad , \quad r_N = \left(\frac{m}{M+m}\right) \tag{1.41}$$

The total angular momentum p_ϕ is

$$p_\phi = M v_N r_N + m v_e r_e = M r_N^2 \omega + m r_e^2 \omega \tag{1.42}$$

or, using Eq.(1.41),

$$p_\phi = \left(\frac{Mm}{M+m} \right) r^2 \omega \tag{1.43}$$

The angular momentum for the two–particle system, Eq.(1.43), is in the same form as for a single particle, Eq.(1.8), except that m is replaced by the reduced mass μ defined by

$$\frac{1}{\mu} = \frac{1}{M} + \frac{1}{m} \quad or \quad \mu = \frac{Mm}{M+m} \tag{1.44}$$

Since M is much larger than m, the reduced mass μ is very close to m, so that the Bohr formula is a very good approximation. Similarly, the total kinetic energy is

$$T = \frac{1}{2} \left(M v_N^2 + m v_e^2 \right) = \frac{1}{2} \left(M r_N^2 + m r_e^2 \right) \omega^2 \tag{1.45}$$

or

$$T = \frac{1}{2} \mu r^2 \omega^2 = \frac{p_\phi^2}{2\mu r^2} \tag{1.46}$$

which is the analog of Eq.(1.9). The centrifugal force on the electron is

$$F = \frac{m v_e^2}{r_e} = m r_e \omega^2 = \frac{p_\phi^2}{\mu r^3}. \tag{1.47}$$

In this way, all of the equations needed for finding the energy E for the planetary model are shown to be identical to those obtained under the assumption of infinite nuclear mass, except that the electron mass m is replaced by μ.

By replacing m by the appropriate value of μ, the Rydberg constant for different nuclear masses can be calculated. If we let R_∞ denote the Rydberg constant for infinite nuclear mass,

$$R_\infty = \frac{me^4}{2\hbar c \hbar^2} = \frac{me^4}{4\pi c \hbar^3} \tag{1.48}$$

while for the hydrogen atom ($\mu = \mu_H$) the Rydberg constant R_H is

$$R_H = \frac{\mu_H e^4}{2hc\hbar^2} = \frac{\mu_H e^4}{4\pi c\hbar^3} \qquad (1.49)$$

The accurate value of R_∞ is 1.0973731×10^5 cm^{-1}. The experimental value for R_H is 1.0967758×10^5 cm^{-1}. The ratio R_∞/R_H is calculated from

$$\frac{R_\infty}{R_H} = \frac{m}{\mu_H} = \frac{m + m_p}{m_p} \qquad (1.50)$$

This means that $R_\infty \geq R_H$. Using the values $m = 9.1091 \times 10^{-28}g$ and $m_p = 1.67252 \times 10^{-24}g$, we find that $R_\infty/R_H = 1.00054$, and the calculated value of R_H is

$$R_H = \left(\frac{m_p}{m + m_p}\right) R_\infty \cong \frac{1.0973731 \times 10^5}{1.00054} = 1.09678 \times 10^5 \ cm^{-1} \quad (1.51)$$

which is in excellent agreement with the experimental value. Correspondingly, the experimentally determined ratio R_H/R_{He} turns out to be the predicted ratio of the reduced masses μ_H/μ_{He}, that is

$$\frac{R_H}{R_{He}} = \frac{\mu_H}{\mu_{He}} \qquad (1.52)$$

Since spectroscopic measurements are very accurate, it is possible to use ratios of Rydberg constants to obtain refined values of relative nuclear masses. Such measurements played a significant role in the discovery of isotopes.

Because of the nuclear mass effect there is an isotopic shift between the spectral lines of different isotopes of the same atom. For example, there is such a shift between the spectrum of atomic deuterium, which has a nucleus with $Z = 1$ but containing a proton and a neutron (so that its mass $M \cong 2m_p$) and that of atomic hydrogen. The ratio of frequencies of corresponding lines is 1.00027, which is detectable, and in fact through this the discovery of the deuteron was made.

Atomic Units:

Quantum mechanical equations are simplified if atomic units (a.u.) are used. In a.u.

Electron mass, $m_e = 1$ $a.u. = 9.109530 \times 10^{-31}$ kg

Electron charge, $e = 1$ $a.u. = 1.602190 \times 10^{-19}$ C

Planck's constant, $\hbar = \frac{h}{2\pi} = 1$ $a.u. = 1.054590 \times 10^{-34}$ Js

Bohr radius, $a_0 = \frac{\hbar^2}{m_e e^2} = 1$ $a.u. = 0.529177 \times 10^{-10}$ m

Fine structure constant, $\alpha = \frac{1}{137.036}$ (dimensionless)

Speed of light, $c = \frac{1}{\alpha} = 137.036$ $a.u. = 2.99792 \times 10^8$ m/s

$\varepsilon_0 \mu_0 = \frac{1}{c^2}$, $\frac{1}{4\pi\varepsilon_0} = 1$ $a.u. = 8.98755 \times 10^9$ m/F

Atomic velocity unit, $v_0 = \alpha c = 1$ $a.u. = 2.18769 \times 10^6$ m/s

Atomic time unit, $t_0 = \frac{a_0}{v_0} = 1$ $a.u. = 2.41889 \times 10^{-17}$ s

Atomic frequency unit, $\nu_0 = \frac{1}{2\pi t_0} = \frac{1}{2\pi}$ $a.u. = 6.57968 \times 10^{15}$ s^{-1}

Atomic energy unit, $E_0 = \frac{e^2}{4\pi\varepsilon_0 a_0} = m_e v_0^2 = 1$ $a.u. = 27.2116$ eV

Radius of Bohr orbit of a one–electron atom, $r_n = \frac{n^2 a_0}{Z} = \frac{n^2}{Z}$ $a.u.$

Bound state energy of a one–electron atom, $E_n = -\frac{m_e Z^2 e^4}{(4\pi\varepsilon_0)^2 2n^2 \hbar^2} = -\frac{Z^2}{2n^2}$ $a.u.$

Atomic unit of electric field, $\mathcal{E}_0 = \frac{e}{a_0^2} = 5.142 \times 10^{11}$ V/m

Atomic unit of magnetic field, $B_0 = 2.35 \times 10^5$ $Tesla$

1.6 Worked examples

Example - 1.1 :

Calculate the ratio of the frequency of radiation emitted in the transition $(n+1) \to n$ and the classical radiation frequency for the orbit n.

Solution :

$$\nu_{n,n+1} = \frac{mZ^2e^4}{2\hbar^2 h}\left[\frac{1}{n^2} - \frac{1}{(n+1)^2}\right] = \frac{mZ^2e^4}{2\hbar^2 h}\left[\frac{2n+1}{n^2(n+1)^2}\right]$$

$$\nu_{class.}^{(n)} = \frac{mZ^2e^4}{2\pi p_\phi^3} = \frac{mZ^2e^4}{2\pi\hbar^3 n^3}$$

$$\frac{\nu_{n,n+1}}{\nu_{class.}^{(n)}} = \frac{(2n+1)n^3}{2n^2(n+1)^2} = \frac{n(2n+1)}{2(n+1)^2} \quad \text{ratio is independent of } Z.$$

Example - 1.2 :

Considering the Bohr model calculate the radius of Li^{++} ion electron in its second excited state.

Solution :

$$r = \frac{n^2\hbar^2}{mZe^2} \quad \longrightarrow \quad \frac{3^3}{3} \ a.u. = 3 \ a.u. = 3 \times 0.529 \ \text{Å} = 1.587 \ \text{Å}.$$

Example - 1.3 :

Numerical calculation of E_1 for H–atom:

Solution :

$$E_n = -\frac{mZ^2e^4}{2\hbar^2 n^2} \text{ for H–atom } Z = 1, \text{ for } n = 1;$$

in a.u. : $E_1 = -\frac{1}{2} \ a.u. = -0.50 \ a.u.$; in SI units:

$$E_1 = -\frac{me^4}{2\hbar^2(4\pi\varepsilon_0)^2} = -\frac{(9.10953\times 10^{-31}kg)\times(1.60219\times 10^{-19}C)^4}{2\times(1.05459\times 10^{-34}Js)^2\times(8.98755\times 10^9 Nm^2/C^2)^{-2}}$$

$E_1 = -2.17990 \times 10^{-18} \ J = -2.17990 \times 10^{-11} \ erg$

$E_1 = -\frac{2.17990 \times 10^{-11}}{1.6021 \times 10^{-12}} \ eV = -13.6065 \ eV \cong -13.6 \ eV.$

Example - 1.4 :

The emission spectrum of the H atom is analyzed between 1000 $\overset{\circ}{A}$ and 4000 $\overset{\circ}{A}$. Which lines are found in this region?

Solution :

Since $\Delta E(cm^{-1}) = R_\infty \left(\frac{1}{n_1^2} - \frac{1}{n_1^2} \right)$, $\lambda(\overset{\circ}{A}) = \frac{1}{\Delta E(cm^{-1})} \frac{1 \overset{\circ}{A}}{10^{-8} \ cm}$

$\lambda(\overset{\circ}{A}) = \frac{10^8}{R_\infty} \frac{n_1^2 n_2^2}{(n_2^2 - n_1^2)}$

For the Lyman series; $n_1 = 1$, $n_2 = 2, 3, \cdots$
$n_1 = 1$, $n_2 = 2$: $\lambda = 1215 \ \overset{\circ}{A}$
$n_1 = 1$, $n_2 = 3$: $\lambda = 1025 \ \overset{\circ}{A}$; these lines will be seen. All other transitions will have a wavelength $< 10^3 \ \overset{\circ}{A}$ and will not be seen.

For the Balmer series; $n_1 = 2$, $n_2 = 3, 4, \cdots$
$n_1 = 2$, $n_2 = 3$: $\lambda = 6561 \ \overset{\circ}{A}$
$n_1 = 2$, $n_2 = 4$: $\lambda = 4860 \ \overset{\circ}{A}$
$n_1 = 2$, $n_2 = 5$: $\lambda = 4339 \ \overset{\circ}{A}$
$n_1 = 2$, $n_2 = 6$: $\lambda = 4101 \ \overset{\circ}{A}$; these lines will not be seen.
$n_1 = 2$, $n_2 = 7$: $\lambda = 3969 \ \overset{\circ}{A}$; will be seen as well as remaining Balmer lines since $n_1 = 2$, $n_2 = \infty$: $\lambda = 3645 \ \overset{\circ}{A}$.

For the Paschen series; $n_1 = 3$, $n_2 = 4, 5, \cdots$
$n_1 = 3$, $n_2 = 4$: $\lambda = 18746 \ \overset{\circ}{A}$
$n_1 = 3$, $n_2 = \infty$: $\lambda = 8201 \ \overset{\circ}{A}$; thus no intervening lines in this series or any later series will be seen.

Example - 1.5 :

The electron with charge $-e$ and the positron with charge $+e$ form a bound state analogous to the hydrogen atom. This system is called positronium. Calculate the ionization potential and the wavelength of the Lyman α line ($n = 1$ to $n = 2$) for positronium.

Solution :

From the Bohr theory we have $\Delta E = R_\infty hc \left(1 + \frac{m_e}{m_{pos}}\right)^{-1} \left(\frac{1}{n_i^2} - \frac{1}{n_f^2}\right)$

For ionization $n_f = \infty$ and $I_p = R_\infty hc \frac{1}{2}$

$I_p = 0.5 \times (1.0973 \times 10^7 \ m^{-1})(6.626 \times 10^{-34} \ Js)(2.998 \times 10^8 \ m/s)$

$= 10.899 \times 10^{-18} \ J = \frac{10.899 \times 10^{-18} \ J}{1.6021 \times 10_{-19} \ J/eV} = 6.80 \ eV$

For the Lyman α line we have:

$\bar{\nu} = \frac{1}{2} R_\infty (1 - \frac{1}{4}) = \frac{3}{8} R_\infty = 4.115 \times 10^6 \ m^{-1}$

The wavelength is therefore

$\lambda = \bar{\nu}^{-1} = 2.430 \times 10^{-7} \ m = 2.430 \ \mathring{A}.$

Example - 1.6 :

Data for the earth–sun system: Mass of the earth $m = 5.983 \times 10^{24} \ kg$, mass of the sun $M = 1.971 \times 10^{30} \ kg$, gravitational constant $G = 6.673 \times 10^{-11} \ Nm^2/kg^2$, earth–sun separation $r = 1.497 \times 10^{11} \ m$.
a) Determine the allowed energy levels for the earth–sun system by quantizing the angular momentum.
b) What is the approximate principal quantum number of the current earth–sun system?
c) What is the energy of the transition $\Delta n = 1$ for the current configu-

ration?

Solution :

a) $\frac{mv^2}{r} = \frac{GmM}{r^2}$, $mvr = n\hbar$ $\rightarrow$ $r = \frac{n^2\hbar^2}{m^2MG}$

$E = T + V = \frac{1}{2}mv^2 - \frac{GmM}{r} = -\frac{GmM}{2r} = -\frac{G^2m^3M^2}{2n^2\hbar^2}$

b) $n = \left(\frac{rm^2MG}{\hbar^2}\right)^{1/2} = 2.52 \times 10^{74}$

c) $\Delta E = E_{n_2} - E_{n_1} = \frac{G^2m^3M^2}{2\hbar^2}\left(\frac{1}{n_1^2} - \frac{1}{n_2^2}\right) \cong \frac{G^2m^3M^2}{2\hbar^2}\left(\frac{1}{n_1^2}\right) \cong 2.08 \times 10^{-41} \ J$

Example - 1.7 :

Assume that the hydrogen atom is held together only by gravitational forces. What would be the radius of this hydrogen atom in the ground state?

Solution :

$r = \frac{n^2\hbar^2}{m_e^2m_pG}$. If $n = 1$ for the lowest state, then

$r = \frac{(1.055\times 10^{-34} \ Js)^2}{(9.108\times 10^{-31} \ kg)^2(1.672\times 10^{-27} \ kg)(6.673\times 10^{-11} \ Nm^2/kg^2)}$

$= 1.201 \times 10^{29} \ m$

(one light year $\cong 9.464 \times 10^{15} \ m$)

Example - 1.8 :

Apply the Wilson–Sommerfeld quantization rule to a particle of mass m that is held to its equilibrium position by a linear restoring force, i.e. $f = -kx$ where k is the force constant and x is the displacement from equilibrium. Obtain an equation for the allowed energies of this simple harmonic oscillator.

Solution :

The potential energy for this system is given by

$$V(x) = -\int_0^x f\,dx = +\tfrac{1}{2}kx^2$$

Therefore, the total energy is

$$E = T + V = \tfrac{p^2}{2m} + \tfrac{kx^2}{2}$$

The maximum amplitude occurs when $p = 0$ so

$$x_{max} = \sqrt{\tfrac{2E}{k}}$$

Finally, the momentum p required in the action integral is obtained

$$p = \sqrt{\left(\tfrac{2E}{k} - x^2\right)km}$$

Then we must evaluate

$$\int p\,dx = 4\int_0^{\sqrt{2E/k}} \sqrt{k}\sqrt{\left(\tfrac{2E}{k} - x^2\right)}\,dx$$

where the factor of 4 is required to account for a complete cycle. The result is

$$\int p\,dx = 2\pi E \sqrt{\tfrac{m}{k}}$$

where we have used integral

$$\int \sqrt{a^2 - x^2}\,dx = \tfrac{1}{2}\left[x\sqrt{a^2 - x^2} + a^2 \sin^{-1}\left(\tfrac{x}{a}\right)\right]$$

obtained by the substitution $x = a\sin\theta$ or from tables. Using $\nu_0 = (1/2\pi)\sqrt{k/m}$ and considering the Wilson–Sommerfeld quantization rule, $\int p\,dx = nh$, we obtain

$$E = nh\nu_0 \quad ; \quad n = 0, 1, 2, \ldots$$

This equation correctly predicts a separation of $h\nu_0$ between the allowed energy levels; however the correct equation is

$$E = (n + \tfrac{1}{2})h\nu_0 \; ; \quad n = 0, 1, 2, ...$$

The important difference between these energies is the "zero point" energy $h\nu_0/2$ which exists with $n = 0$.

Example - 1.9 :

When the nucleus in a hydrogen–like atom is assumed to be fixed in space the allowed energies are given by $E_n = -hcR_\infty/n^2$. Derive the correction factor that must be used to account for the finite mass of the nucleus.

Solution :

When the electron mass is m and the nuclear mass is M the rotational kinetic energy of the Bohr atom is given by $T = \tfrac{1}{2}\mu v^2$ where $\mu = mM/(m + M)$. Therefore, m should be replaced in the final energy equation by μ to give

$$E_n = -hcR_\infty(\frac{M}{m + M})/n^2 = -hcR_\infty(1 + \frac{m}{M})^{-1}/n^2$$

Example - 1.9 :

What is the electric field–strength of the nucleus for the first and fourth Bohr orbits of the hydrogen atom?

Solution :

$E = \frac{e}{a^2} \times 300 \; V/cm$, where a is the orbit radius.

$E_1 = 5.13 \times 10^9 \; V/cm$, $E_4 = 2 \times 10^7 \; V/cm$

Chapter 2

Radiation and Matter

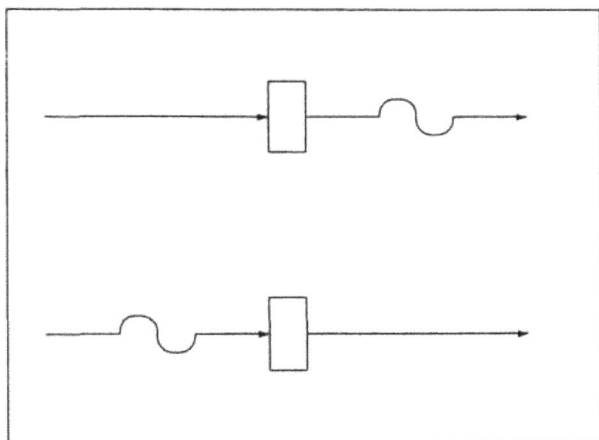

2.1   The nature of radiation and matter

It had been assumed that matter is distinct from radiation, which is the transmission of energy by wave motion. However, experiments performed in the period 1887–1927 demonstrated that the boundary between matter with its particle–like behaviour, and radiation with its wave–like behaviour, is not as rigid as has been supposed. In fact, the observations implied that both matter and radiation can behave as if they are composed of waves or as if they are composed of particles, the behaviour manifested depending on the nature of the experiment.

2.2   The particle–like character of radiation

Bohr's model for radiation by the hydrogen atom provides a hint that radiant energy can have a particle–like character as well. Since radiation was assumed by Bohr to be emitted or absorbed in indivisible units or quanta, $h\nu$, it is reasonable to suppose that these quanta might behave like particles. To understand some properties of the radiation particles, called photons, we make use of the theory of relativity.

If a particle has (rest) mass m and velocity v, special relativity theory gives the momentum p as

$$p = mv \times \left(1 - \frac{v^2}{c^2}\right)^{-\frac{1}{2}} \tag{2.1}$$

In order to have a finite momentum, photons must have zero mass. The energy E of a relativistic particle of mass m moving freely in space with momentum p is

$$E = \left(m^2 c^4 + p^2 c^2\right)^{\frac{1}{2}} \tag{2.2}$$

For a photon, $E = h\nu$ and $m = 0$, so that

$$p = \frac{E}{c} = \frac{h\nu}{c} \tag{2.3}$$

This formula for the photon momentum characterizes one of the particle aspects of radiation.

We now examine some experimental results which cannot be understood in terms of the wave theory.

2.3　The photoelectric effect

If light is allowed to impinge upon the clean surface of a metal in an evacuated vessel, electrons are liberated from the metal. The photoelectrons can be collected by a plate and a circuit set up to measure the current.

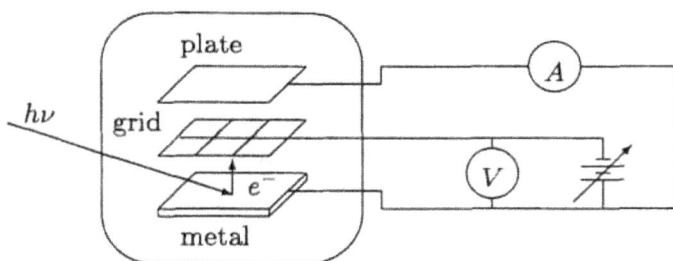

Fig. 2.1. Schematic diagram for photoelectric experiment.

The photoelectric current, which is the amount of charge arriving at the plate per unit time, is proportional to the rate of emission of electrons from the metal surface,

$$\frac{\Delta n}{\Delta t} = \frac{i}{e} \quad , \quad i = \frac{\Delta n e}{\Delta t} = \frac{\Delta q}{\Delta t} \tag{2.4}$$

To determine the velocity with which the photoelectrons travel, a potential is applied to a grid mounted between the metal surface and the plate. The potential creates an electric field which decelerates the photoelectrons. Let the stopping voltage is V_s. At the stopping voltage, the initial kinetic energy of photoelectrons has been converted to potential energy.

$$\frac{1}{2}mv^2 = eV_s \tag{2.5}$$

Thus, by measuring i and V_s we know the number of electrons produced per second and their maximum kinetic energy.

The experimental result shows that V_s is proportional to the frequency of the light and independent of the intensity. If the frequency ν is below a certain threshold value ν_0, no photoelectron current is produced. At frequencies greater than ν_0, the empirical equation for the stopping voltage is

$$V_s = k(\nu - \nu_0) \tag{2.6}$$

where k is a constant independent of the metal used, but ν_0 varies from one metal to another. Although there is no relation between V_s and the light intensity, I, it is found that the photoelectric current, and therefore the number of electrons liberated per second, is proportional to I. These results can not be explained by the wave theory. However, in the particle model, a photon of energy $h\nu$ strikes a bound electron, which may be absorb the photon energy. If $h\nu$ is greater than the binding energy (or work function) eV_0, the electron is liberated. Thus, the threshold frequency ν_0 is given by

$$\nu_0 = \frac{eV_0}{h} \tag{2.7}$$

Since V_0 is a characteristic of the particular metal, ν_0 depends upon the metal. For a photon of energy $h\nu$, the total energy of the struck electron is $h\nu$, with the excess over the potential energy eV_0 required to escape from the metal appearing as kinetic energy,

$$\frac{1}{2}mv^2 = h\nu - eV_0 = eV_s \tag{2.8}$$

which is identical to the empirical relationship Eq.(2.6), with $k = h/e$.

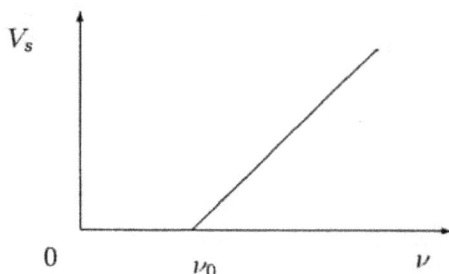

Fig. 2.2. V_s vs ν in photoelectric measurements.

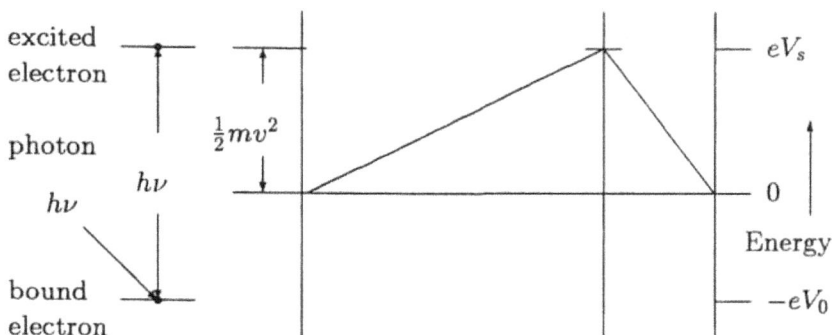

Fig. 2.3. Potential energy of an electron.

Because the amount of energy absorbed by an electron is $h\nu$ regardless of the rate at which photons impinge on the surface, the kinetic energy of the ejected electrons should be independent of the intensity of the light. So this experiment demonstrates the particle character of radiation.

2.4 The Compton effect

The Compton experiment provides detailed information about the interaction of radiation and matter which was performed and analyzed by A.H. Compton in 1923.

A sample of material is irradiated (for example, paraffin and X-rays). The wavelength of the scattered radiation and the energy of the emitted electron are determined as a function of angle relative to the incident beam. It is found that the radiation scattered from the material contains the radiation having the same wavelenght λ with the incident radiation and radiation with longer wavelength λ'. The relation between λ and λ' is found to be

$$\lambda' = \lambda + k \sin^2(\theta/2) \tag{2.9}$$

where k is a constant. The classical theory does not correctly explain the observations. The observations can be explained quantitatively by

the photon theory of radiation and the laws of conservation of energy
and momentum for particles.

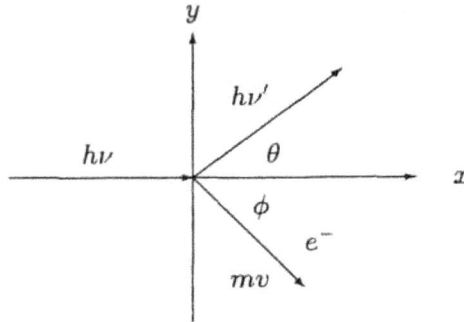

Fig. 2.4. The Compton effect.

The incident photon with wavelength λ and frequency $\nu = c/\lambda$ has
momentum $p = h\nu/c$. The scattered photon which has a longer wavelength λ', and therefore a lower frequency $\nu' = c/\lambda'$ has a lower momentum $p' = h\nu'/c$. The electron is ejected in the direction ϕ with a
momentum mv.

Conservation equations: For energy

$$h\nu = h\nu' + \frac{1}{2}mv^2 \tag{2.10}$$

for momentum in the x-direction

$$\frac{h\nu}{c} = \frac{h\nu'}{c}\cos(\theta) + mv\cos(\phi) \tag{2.11}$$

for momentum in the y-direction

$$0 = \frac{h\nu'}{c}\sin(\theta) + mv\sin(\phi) \tag{2.12}$$

Eliminating v and ϕ from the equations and making the approximation
that $\lambda\lambda' \cong \lambda^2$, we find that

$$\Delta\lambda = \lambda' - \lambda = 2\left(\frac{h}{mc}\right)\sin^2(\theta/2) \tag{2.13}$$

$$= \left(\frac{h}{mc}\right)(1 - \cos\theta) = \lambda_c(1 - \cos\theta)$$

where λ_c is called the Compton wavelength of the particle with mass m. For electron it is 0.02425 Å. For λ in Å Eq.(2.13) gives

$$\Delta\lambda = 0.0485\sin^2(\theta/2) \quad \text{Å} \tag{2.14}$$

The maximum shift is for $\theta = \pi$, where $\Delta\lambda = 0.0485$ Å.

The particle point of view and Newtonian mechanics lead to a simple and quantitatively correct interpretation of these experiments, and that predictions based upon the classical wave theory are wrong. However, for diffraction and interference phenomena the wave theory works well and the particle theory does not. Therefore, the character exhibited by radiation, whether wave–like or particle–like, depends upon the type of experiment.

If the interaction of radiation with matter produces a measurable change in the matter, such as the ejection of an electron, the phenomenon appears to require the photon theory. If the interaction produces a measurable change in the spatial distribution of the radiation, such as diffraction, but produces no measurable change in the matter, the wave theory must be considered. These results suggest that the measurement process itself must be included in the theory.

2.5   The wave nature of matter

The dual nature of radiation applies equally to matter. The duality of matter was suggested on the basis of theoretical considerations prior to its experimental verification.

For a photon, the momentum and wavelength are simply related, $p = h/\lambda$. In 1924, de Broglie reasoned that a similar relation should hold for material particles, since the same relativistic equations of motion apply to photons and to particles of nonzero rest mass. Thus, an electron with momentum p should have associated with it a wave whose wavelength is given by

$$\lambda = \frac{h}{p} \tag{2.15}$$

For a simple calculation to estimate the wavelength of a moving electron, consider an electron accelerated through a potential V. The kinetic energy of electron is $E = mv^2/2 = eV$, and the momentum is $p = mv = \sqrt{2meV}$. Using Eq.(2.15), we find that

$$\lambda = \frac{h}{p} = \frac{h}{\sqrt{2meV}} = \frac{7.083 \times 10^{-9}}{\sqrt{V}} \quad cm \qquad (2.16)$$

in cgs units $h/\sqrt{2me} = 6.6262 \times 10^{-27}/(2 \times 9.10956 \times 10^{-28} \times 4.80325 \times 10^{-10})^{1/2} = 7.08326 \times 10^{-9}$. Here V is in statvolts (1 *statvolt* $= 300\ V$). For a potential of 1 statvolt, $\lambda \cong 10^{-8} cm = 1\ $Å. This is a very small length to observe the diffraction of electrons in an ordinary apparatus. To perform an experimental test of de Broglie's hypothesis, we must have gratings or slits with spacing on the order of 1Å. Suitable natural gratings are furnished by the array of atoms in a crystal.

2.6   Diffraction of electrons

C. Davisson and L.H. Germer in 1927 performed the first experiment of diffraction patterns for electrons reflected from a crystal.

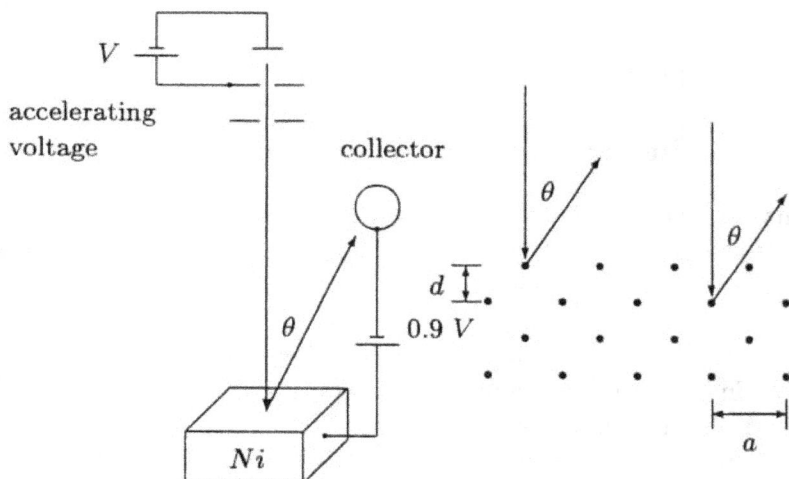

Fig. 2.5. Schematic diagram of Davisson–Germer experiment.

The collector measures the electron current as a function of θ. Stopping voltage (which stops electrons which have lost more than 10% of their kinetic energy) restricts the measurements to elastically scattered electrons.

By the wave theory, if λ is the wavelength of a wave associated with the electron beam, constructive interference occurs for angles at which the waves scattered by different atoms have paths differing in length by $n\lambda$ (n is an integer); if the path lengths differ by an amount $n\lambda/2$ (n is odd), destructive interference occurs.

Davisson and Germer obtained plots of the electron current I versus θ as shown in Fig. 2.6. The appearance of a maximum in the I vs θ plot shows that the electron beam is scattered preferentially at a certain angle.

For electron beams with the low velocities used by Davisson & Germer ($V \cong 10 - 400\ V$), the penetration of the electrons into the crystal is very slight, so that the diffraction occurs due to the electrons reflected from the surface layer.

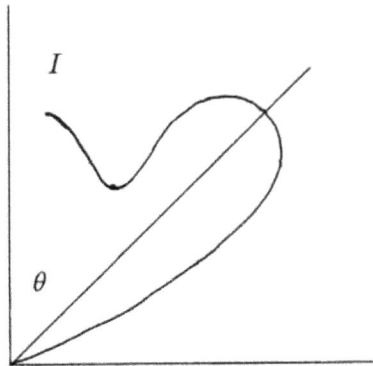

Fig. 2.6. Polar plot of I vs θ.

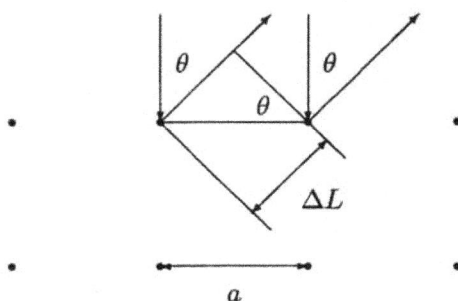

Fig. 2.7. Single–plane diffraction model.

The path length difference ΔL is $a\sin(\theta)$. If this distance is an integer number of wavelength, the waves are in phase and an intensity maximum is observed at θ. Thus, the law for single–plane diffraction is

$$n\lambda = a\sin(\theta) \qquad (2.17)$$

Davisson and Germer found that for $V = 0.18$ statvolt the maximum scattering angle for nickel is at $\theta = 50.7$ deg. $a = 2.15$ Å for Ni. The wavelength calculated from Eq.(2.17) is 1.66 Å. According to Eq.(2.16) $\lambda = 7.083 \times 10^{-9}/\sqrt{0.18}$ $cm = 1.67$ Å in good agreement with Davisson and Germer's result.

G.P. Thomson also studied electron diffraction patterns. He took very thin metal films as target, and the electron energies were sufficient to penetrate the films. The beam of electrons passes through the metal film.

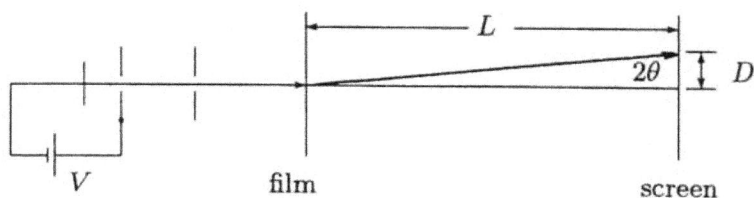

Fig. 2.8. Schematic diagram of Thomson experiment.

Since the target is penetrated, we must consider the reflections from inner layers. If the deflection angle (2θ) is one of the maximum intensity, the difference in path length between the rays reflected from the different planes must be an integral multiple of the wavelength.

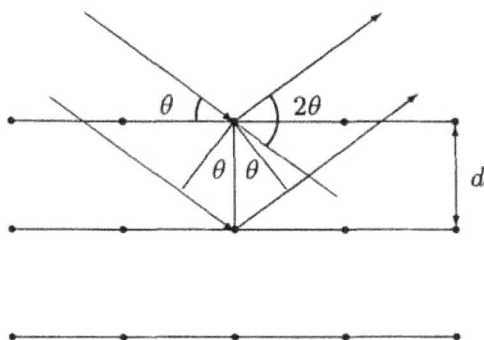

Fig. 2.9. Double–plane diffraction model.

For two adjacent planes this difference is $2d\sin(\theta)$; thus the intensity maxima occur for angles which satisfy Bragg's law,

$$n\lambda = 2d\sin(\theta). \tag{2.18}$$

The distance D from the origin to the center of an incident ring is related to the deflection angle 2θ and the distance L from the film to the screen by

$$\frac{D}{L} = \tan(2\theta) \tag{2.19}$$

Since in Thomson's apparatus $L \gg D$, θ is small and $D/L = \tan(2\theta) \cong 2\sin(\theta)$. Using Eq.(2.18) we obtain

$$d = \frac{n\lambda L}{D}. \tag{2.20}$$

If a high energy electron beam is passed through a gas, diffraction patterns similar to those for metal films are the result.

Observations of diffraction patterns for beams of particles have shown that the de Broglie relation is generally applicable and that all particles behave in a wave–like manner under appropriate experimental conditions.

2.7   De Broglie formula and the H–atom

It is assumed that for a stationary state of the hydrogen atom, the wave associated with the electron orbit must be a stationary wave; that is, the positions of its maxima and minima do not change with time.

For this to occur, there must be an integral number of wavelength in the circular orbit of the electron.

$$2\pi r = n\lambda \quad , \quad n = 1, 2, 3, ... \tag{2.21}$$

Stationary ($2\pi r = n\lambda$) and non–stationary ($2\pi r \neq n\lambda$) waves:

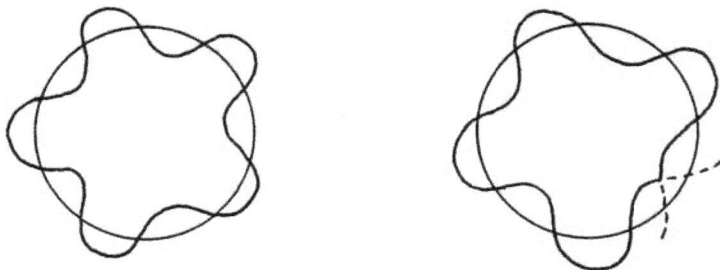

Applying the de Broglie relation, we find that

$$2\pi r = n\lambda = n\frac{h}{p} = \frac{nh}{mv} = \frac{nhr}{p_\phi} \tag{2.22}$$

here we used $p_\phi = mvr$ for the angular momentum. Elimination of r from Eq.(2.22) leads directly to the quantization rule for angular momentum, $p_\phi = n\hbar$. These qualitative ideas were incorporated into the systematic theory of quantum mechanics.

2.8   Uncertainty principle and the Bohr atom

According to wave mechanics the possible uncertainty relations are:

$$\Delta x \Delta p \geq h \quad , \quad \Delta E \Delta t \geq h \quad , \quad and \quad \Delta k \Delta x \geq 1 \qquad (2.23)$$

Some textbooks take these inequalities as $\geq \hbar/2$ or $\geq \hbar$. The uncertainty principle shows us why classical theory is not applicable to phenomena on a microscopic scale.

Let us consider an experiment to measure accurately the Bohr radius a_0. We imagine that we can observe the position of the electron in a hydrogen atom in its ground state by looking at it through a super microscope of some kind. A photon is expected to undergo Compton scattering by the electron in the Bohr orbit and suffer a shift in wavelength on the order of $\Delta \lambda \cong h/mc$.

The energy gained by the electron in the scattering is

$$\Delta E = h \Delta \nu = hc \left(\frac{1}{\lambda'} - \frac{1}{\lambda} \right) \cong \frac{hc\Delta\lambda}{\lambda^2} \cong \frac{h^2}{\lambda^2 m} \qquad (2.24)$$

The magnitude of the radial component of the final momentum of the electron in its orbit must be between the limits

$$0 \leq p_r \leq \sqrt{2m\Delta E} \qquad (2.25)$$

Therefore, the maximum uncertainty in p_r is

$$\Delta p_r = \sqrt{2m\Delta E} = \sqrt{2}\frac{h}{\lambda} \qquad (2.26)$$

Since we wish the atom to remain in its ground state during the measurement, we must choose λ so large that ΔE is less than the energy required for excitation to the $n = 2$ state, that is,

$$\Delta E \cong \frac{h^2}{\lambda^2 m} < \frac{e^2}{2a_0} \left(\frac{1}{1^2} - \frac{1}{2^2} \right) = \frac{3e^2}{8a_0} \qquad (2.27)$$

Solving Eq.(2.27) for λ, we have

$$\lambda > \left(\frac{8h^2 a_0}{3me^2} \right)^{1/2} = \sqrt{2/34\pi a_0} \quad ; \quad \left(a_0 = \frac{\hbar^2}{me^2} \right) \qquad (2.28)$$

We can calculate the uncertainty in the radius of the electron from the uncertainty in momentum,

$$\Delta r = \frac{h}{\Delta p_r} \cong \frac{\lambda}{\sqrt{2}} \tag{2.29}$$

or

$$\Delta r > \frac{4\pi}{\sqrt{3}} a_0 \tag{2.30}$$

Eq.(2.30) shows that a direct measurement of a_0 has resulted in a minimum uncertainty in the position of the electron which is larger than the magnitude of a_0 itself. Thus, according to the uncertainty principle, Bohr's picture of fixed radii for electron orbit is not realistic.

If there is an inherent indeterminacy in a certain measurement, then we speak only of the likelihood or probability that the experiment will yield certain results. Wave mechanics provides a method by which these probabilities can be calculated. Newtonian mechanics is a limiting theory for macroscopic dimensions, the quantum theory is an approximation that is valid only above certain subatomic dimensions.

2.9   Worked examples

Example - 2.1 :

When a clean tungsten surface is illuminated by light of wavelength 200 nm, a potential of 1.68 V is required to stop the emitted electrons. When the wavelength of light is 150 nm, a potential of 3.74 V is required. From these data determine the Planck's constant (in $erg \cdot s$) and the potential barrier to emission (in eV) for tungsten.

Solution :

$$eV_1 = h\nu_1 - W \quad , \quad eV_2 = h\nu_2 - W$$

$$e(V_1 - V_2) = h(\nu_1 - \nu_2)$$

$$h = \frac{e(V_1 - V_2)}{\nu_1 - \nu_2)} = \frac{e}{c} \frac{V_1 - V_2}{\frac{1}{\lambda_1} - \frac{1}{\lambda_2}} = \frac{e}{c} \lambda_1 \lambda_2 \left(\frac{V_1 - V_2}{\lambda_2 - \lambda_1} \right)$$

$$\cong \frac{1.6\times10^{-19}}{3\times10^8} \times 1.5 \times 10^{-7} \times 2.0 \times 10^{-7} \times \frac{(1.68-3.75)}{(1.5-2.0)\times10^{-7}} \quad (Js)$$

$$\cong 6.592 \times 10^{-34} \quad Js = 6.592 \times 10^{-27} \quad erg\ s$$

$$W = h\nu_1 - eV_1 \cong 6.592 \times 10^{-34} \times \frac{3\times10^8}{2\times10^{-7}} - 1.6 \times 10^{-19} \times 1.68 \quad (J)$$

$$= (6.592 \times 1.5 - 1.6 \times 1.68) \times 10^{-19}/1.6 \times 10^{-19} \quad eV \cong 4.5\ eV.$$

Example - 2.2 :

In a Compton experiment the deflection of incident photon is measured to be 60^0. Calculate the shift in wavelength of the photon. Assume photon interacts with a single electron.

Solution :

$$\Delta\lambda \cong 2\left(\frac{h}{mc}\right)\sin^2(\theta/2) \cong 0.0485 \times \sin^2(\theta/2) \quad \mathring{A}$$

$$= 0.0485 \times (1/2)^2 \quad \mathring{A} = 0.0485/4 \quad \mathring{A} \cong 0.0121\ \mathring{A}$$

Example - 2.3 :

A photon of initial wavelength $0.4\ \mathring{A}$ suffers two successive collisions with two electrons. The deflection in the first collision is 90^0 and in the second collision it is 60^0. What is the final wavelength of the photon?

Solution :

First: $\Delta\lambda_1 = \left(\frac{h}{mc}\right)\left[1 - \cos(90^0)\right] = 2.42 \times 10^{-12}\ m$

Second: $\Delta\lambda_2 = \left(\frac{h}{mc}\right)\left[1 - \cos(60^0)\right] = 1.21 \times 10^{-12}\ m$

Therefore: $\Delta\lambda = \Delta\lambda_1 + \Delta\lambda_2 = 3.63 \times 10^{-12}\ m = 0.036\ \mathring{A}$

Final wavelength : $\lambda = 0.4 + 0.036 = 0.436$ Å.

Example - 2.4 :

Find the ratio of the Compton wavelength λ_c to the de Broglie wavelength λ for a relativistic electron.

Solution :

$$\lambda_c = \frac{h}{mc} \quad , \quad \lambda = \frac{h}{p} \quad ; \quad \frac{\lambda_c}{\lambda} = \frac{h/mc}{h/p} = \frac{p}{mc}$$

$$E^2 = c^2 p^2 + (mc^2)^2 \quad \longrightarrow \quad p = \sqrt{\frac{E^2}{c^2} - (mc)^2}$$

$$\frac{\lambda_c}{\lambda} = \frac{1}{mc}\sqrt{\frac{E^2}{c^2} - (mc)^2} = \sqrt{\left(\frac{E}{mc^2}\right)^2 - 1} \; .$$

Example - 2.5 :

Calculate the de Broglie wavelength of
a) a photon with energy 1 eV, 100 eV
b) an electron with kinetic energy 1 eV, 100 eV
c) a proton with kinetic energy 1 eV
d) a UF_6 molecule with kinetic energy 1 eV
e) a baseball with velocity 40 m/s (mass = 0.1 kg).

Solution :

a) $E = h\nu = h\frac{c}{\lambda} \quad \rightarrow \quad \lambda = \frac{hc}{E}$

1 eV photon : $\lambda = \frac{(6.626\times 10^{-34}\ Js)(2.998\times 10^{8}\ m/s)}{(1\ eV)(1.602\times 10^{-19}\ J/eV)} \cong 1.24 \times 10^{-8}\ m$

100 eV photon : $\lambda = \frac{1.24\times 10^{-8}}{100} = 1.24 \times 10^{-10}\ m$

b) For particles with $v \ll c$, $E = \frac{p^2}{2m} = \frac{h^2}{2m\lambda^2}$

$\lambda = \frac{h}{\sqrt{2mE}} = \frac{1.170\times 10^{-24}}{\sqrt{m(kg)E(eV)}} \quad (in\ m)$

1 eV electron : $\lambda = 1.226 \times 10^{-9}$ m

100 eV electron : $\lambda = 1.226 \times 10^{-11}$ m

c) 1 eV proton : $\lambda = 2.862 \times 10^{-11}$ m

d) 1 eV UF_6 (mass $= 5.855 \times 10^{-25}$ kg/molecule) : $\lambda = 1.529 \times 10^{-12}$ m

e) $\lambda = \frac{h}{mv} = \frac{6.626 \times 10^{-34}\ Js}{(0.1\ kg)(40\ m/s)} \cong 1.656 \times 10^{-34}$ m.

Example - 2.6 :

Consider the diffraction of a beam of thermalized He atoms by a simple cubic lattice ($d \cong 2\ \mathring{A}$). At what temperature for the He atoms would diffraction be appreciable.

Solution :

$$\lambda = \frac{h}{mv} = \frac{h}{p}$$

If the He atoms are thermalized, then the kinetic theory of gases predicts that the average kinetic energy of the atoms will be $E = \frac{3}{2}k_BT$. Thus

$$E = \frac{3}{2}k_BT = \frac{1}{2}mv^2 = \frac{p^2}{2m} = \frac{h^2}{2m\lambda^2}$$

or $T = \frac{h^2}{3mk_B\lambda^2}$

For the diffracted beam to have a reasonable intensity, the lattice spacing d and the wavelength of the radiation must be roughly of the same magnitude. Thus

$$T \cong \frac{(6.626 \times 10^{-34}\ Js)^2 \times (6.023 \times 10^{23}\ mol^{-1})}{3 \times (4.0026 \times 10^{-3}\ kg/mol) \times (1.38 \times 10^{-23}\ J/K) \times (2 \times 10^{-10}\ m)^2}$$

$T \cong 39\ K$.

Example - 2.7 :

When a charged particle moves rapidly in a region of space filled by isotropic electromagnetic radiation (for instance, light from the Sun and stars), the particle loses energy as a result of the interaction with the radiation (assuming that the energy of the particle is greater than the energy of the photons which make up the radiation). Assuming that the particle is ultrarelativistic (its energy $E_0 \gg mc^2$) and that is collides head–on with the photon, find the particle energy change $E_0 - E = \Delta E$ and the energy of recoil $\hbar\omega$ of the photon. Assume that the energy $\hbar\omega_0$ of the photon (before collision) is small compared with $\hbar\omega$. Analyse the result. What is the energy $\hbar\omega$, if the particle is an electron with energy $E_0 = 2.5 \times 10^9$ eV and $\hbar\omega_0 = 1$ eV.

Solution :

$$\Delta E = \hbar\omega = \left(\frac{2E}{mc^2}\right) \hbar\omega_0 / \left(1 + 4\frac{E_0\hbar\omega_0}{m^2c^4}\right) . \text{ If } E_0 \ll mc^2(mc^2/4\hbar\omega_0),$$

then $\hbar\omega \cong (2E/mc^2)^2\hbar\omega_0 \ll E_0$; in the opposite limiting case $\hbar\omega \cong E_0$. In the example given, $\hbar\omega = 10^8$ eV.

Example - 2.8 :

Find an expression for the refractive index of electronic waves in terms of the work function $U_0 = eV_0$ (V_0 is the internal potential of the crystal).

Solution :

$\mu = \frac{\lambda_0}{\lambda} = \sqrt{\frac{E+U_0}{E}} = \sqrt{1 + \frac{V_0}{V}}$. Here E and V are respectively the electron energy in vacuo and the potential difference corresponding to it.

Chapter 3

Wave Equations for Simple Quantum Systems

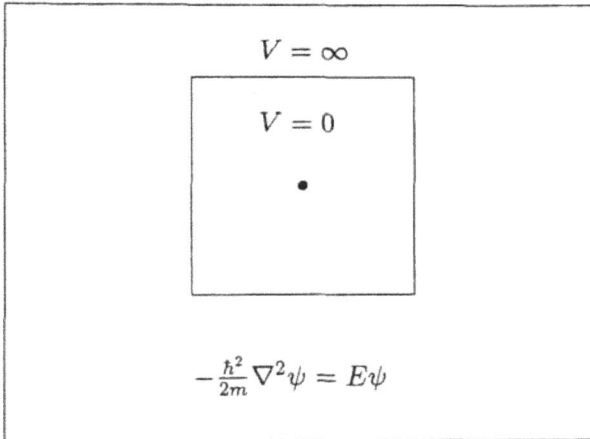

$$V = \infty$$

$$V = 0$$

•

$$-\frac{\hbar^2}{2m}\nabla^2\psi = E\psi$$

3.1   The Schrödinger equation

The basic equation of quantum mechanics is known as the Schrödinger equation. The equation is a postulate and therefore it is not possible to prove that the equation must be true based on first principles. The Schrödinger equation describes non–relativistic particles, whose energy E and momentum p are related by

$$E = \frac{p^2}{2m} \tag{3.1}$$

Non–relativistic kinematics can be used so long as the energy E is not comparable with, or larger than, the rest–mass energy mc^2.

According to the rules of quantum mechanics states of a system are described by wave functions (Ψ), observable quantities are associated with operators ($\hat{O}$), and when the value of an observable O is known to be o, the system is in a state whose wave function is an eigenfunction of the operator $\hat{O}$ corresponding to O, with eigenvalue o. That is

$$\hat{O}\Psi = o\Psi \tag{3.2}$$

In general such an equation is called as eigenvalue equation. If the eigenvalue o describes the energy (E) of a system then the corresponding operator $\hat{O}$ is called the Hamiltonian operator or simply Hamiltonian of the same system. Then the corresponding eigenvalue equation is expressed as

$$H\Psi = E\Psi \tag{3.3}$$

If the system described by such an eigenvalue equation is in a state such that its energy takes a definite value E then the eigenvalue equation is called as the time–independent Schrödinger equation, Eq.(3.3). On the other hand, if the Hamiltonian depends explicitly on the time t as well as the position $\mathbf{r}$ of the system, then the solution (or the wave function which describes the system) depends on time also. In this case the corresponding eigenvalue equation is expressed as

$$H\Psi = i\hbar \frac{d\Psi}{dt} \tag{3.4}$$

This is the time–dependent Schrödinger equation.

3.2   The free particle wave equation in one dimension

Time–independent stationary wave for a free particle may be written as

$$\psi(x) = \psi_0 \cos(2\pi x/\lambda) \tag{3.5}$$

where ψ_0 is the amplitude of the wave.

The classical time–independent wave equation is

$$\frac{d^2\psi(x)}{dx^2} = -\psi_0(2\pi/\lambda)^2 \cos(2\pi x/\lambda) = -(2\pi/\lambda)^2\psi(x) \tag{3.6}$$

using the de Broglie relation one writes

$$\frac{d^2\psi(x)}{dx^2} = -\frac{p^2}{\hbar^2}\psi(x) \tag{3.7}$$

Since the energy E of a free particle is entirely kinetic energy, we can write

$$E = \frac{1}{2}mv^2 = \frac{p^2}{2m} \tag{3.8}$$

or

$$p^2 = 2mE \tag{3.9}$$

The wave equation, Eq.(3.7), then takes the form

$$\frac{d^2\psi(x)}{dx^2} = -\frac{2mE}{\hbar^2}\psi(x) \quad or \quad -\frac{\hbar^2}{2m}\frac{d^2\psi(x)}{dx^2} = E\psi(x) \tag{3.10}$$

Eq.(3.10) is called the Schrödinger amplitude equation for a free particle. Defining

$$k^2 = \frac{2mE}{\hbar^2} \tag{3.11}$$

Eq.(3.10) can be written

$$\frac{d^2\psi(x)}{dx^2} = -k^2\psi(x) \tag{3.12}$$

The general solution to this second order ordinary differential equation is

$$\psi(x) = Ae^{ikx} + Be^{-ikx} \quad or \quad \psi(x) = A\cos(kx) + B\sin(kx) \tag{3.13}$$

The corresponding energy is continuous

$$E = \frac{\hbar^2 k^2}{2m} \tag{3.14}$$

The trouble with these wave functions is that they are not square integrable, since $\int_{-\infty}^{\infty} |\psi(x)|^2 dx$ diverges for all values of A and B. One way of avoiding the normalization difficulty is to deal with the probability current, or **flux**

$$j(x) = \frac{\hbar}{2im} \left[\psi^*(x) \frac{d\psi(x)}{dx} - \frac{d\psi^*(x)}{dx} \psi(x) \right] = v_z(A^2 - B^2) \tag{3.15}$$

3.3   The free particle wave equation in three dimensions

Each squared component of momentum obey an equation equivalent to Eq.(3.7),

$$-\hbar^2 \frac{\partial^2 \psi(x, y, z)}{\partial x^2} = p_x^2 \psi(x, y, z) \tag{3.16}$$

$$-\hbar^2 \frac{\partial^2 \psi(x, y, z)}{\partial y^2} = p_y^2 \psi(x, y, z) \tag{3.17}$$

$$-\hbar^2 \frac{\partial^2 \psi(x, y, z)}{\partial z^2} = p_z^2 \psi(x, y, z) \tag{3.18}$$

As in the one–dimensional case, all of the energy is kinetic energy, we have

$$p^2 = p_x^2 + p_y^2 + p_z^2 = \hbar^2(k_x^2 + k_y^2 + k_z^2) = 2mE \tag{3.19}$$

Adding Eqs.(3.16–18) and using Eq.(3.19) we obtain

$$-\frac{\hbar^2}{2m} \left(\frac{\partial^2}{\partial x^2} + \frac{\partial^2}{\partial y^2} + \frac{\partial^2}{\partial z^2} \right) \psi(x, y, z) = E\psi(x, y, z) \tag{3.20}$$

The probability of finding the particle in the differential volume element $dxdydz$ between x and $x + dx$, y and $y + dy$, z and $z + dz$ is given by

$$\int \int \int \psi^*(x, y, z)\psi(x, y, z)dxdydz \quad or \quad \int |\psi(\mathbf{r})|^2 d\mathbf{r} \tag{3.21}$$

The general solution looks like the one–dimensional case,

$$\psi(\mathbf{r}) = A e^{i\mathbf{k}\cdot\mathbf{r}} + B e^{-i\mathbf{k}\cdot\mathbf{r}} \qquad (3.22)$$

Since normalization is not possible for this wave function, the flux for this case is expressed as

$$\mathbf{j}(\mathbf{r}) = \frac{\hbar}{2im} \left[\psi^*(\mathbf{r}) \nabla \psi(\mathbf{r}) - \nabla \psi^*(\mathbf{r}) \psi(\mathbf{r}) \right] = \mathbf{v}(A^2 - B^2) \qquad (3.23)$$

3.4 Particle in a one–dimensional box

A particle of mass m in a one–dimensional infinite square well potential energy box of length L can be represented by the wave equation

$$-\frac{\hbar^2}{2m} \frac{d^2\psi}{dx^2} = E\psi \quad or \quad \frac{d^2\psi}{dx^2} + k^2\psi = 0 \quad ; \quad k^2 = \frac{2mE}{\hbar^2} \qquad (3.24)$$

In this case the wave function obey the boundary conditions

$$\psi(0) = \psi(L) = 0 \qquad (3.25)$$

Then the wave function takes the general form

$$\psi(x) = A\sin(kx) + B\cos(kx) \quad , \quad 0 \le x \le L \qquad (3.26)$$

Since $\cos(kx)$ is an even function of x and $\sin(kx)$ is an odd function of x, the even parity eigenfunctions are

$$\psi(x) = B\cos(kx) \quad , \quad 0 \le x \le L \qquad (3.27)$$

and the odd parity eigenfunctions are

$$\psi(x) = A\sin(kx) \quad , \quad 0 \le x \le L \qquad (3.28)$$

In both cases the values of k must be such as to satisfy condition (3.25). Because of the evenness or oddness of the eigenfunctions, it is only necessary to apply condition (3.25) at $x = L$. From Eq.(3.27) we have $0 = B\cos(kL)$, this demands $kL = n\pi/2$; $n = 1, 3, 5, \cdots$ From Eq.(3.28) we have $0 = A\sin(kL)$, this demands $kL = n\pi$; $n = 1, 2, 3, \cdots$ We may take the odd parity functions as the solution to the wave equation,

Eq.(3.24), because they are defined for all positive integer values of n.

The corresponding energy values are

$$E_n = \frac{n^2 h^2}{8mL^2} \quad , \quad n = 1, 2, 3, ... \tag{3.29}$$

Energy is quantized.

$\psi^2(x)$ is interpreted as the probability per unit volume or the probability density for the particle. The quantity

$$\psi^2(x)dx \tag{3.30}$$

is the probability of finding the particle in the range x to $x + dx$.

Integration of Eq.(3.30) over all values of x must give unity, which is the nornalization condition

$$\int_0^L \psi^2(x)dx = 1 \tag{3.31}$$

The normalized wave function has the form

$$\psi_n(x) = \sqrt{2/L}\sin(n\pi x/L) \tag{3.32}$$

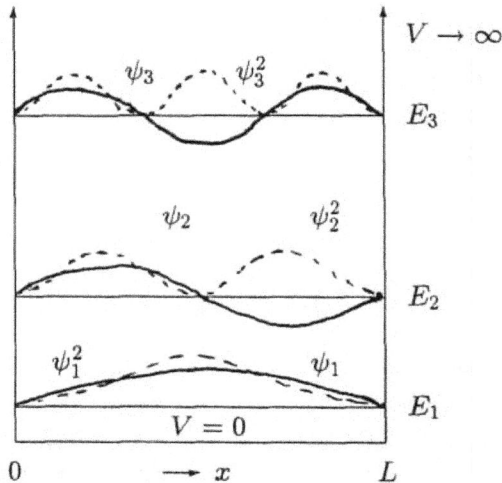

Fig. 3.1. Energy levels of a particle in infinite well.

For a system in a given stationary state, the energy is known exactly from the solution of the Schrödinger equation. However, as suggested by the uncertainty principle, the momentum, like the position, is usually not completely specified, and average or expectation values are of interest, which are defined as

$$< x > = \int \psi^*(x) x \psi(x) dx \qquad (3.33)$$

$$< p > = \int \psi^*(x) p \psi(x) dx \qquad \left(p \rightarrow -i\hbar \frac{d}{dx} \right) \qquad (3.34)$$

For a particle in a box

$$< x > = L/2 \quad , \quad < p > = 0 \qquad (3.35)$$

3.5   Particle in a three–dimensional box

For a particle in a three–dimensional box of sides a, b, and c the resulting normalized wave functions and the associated values of the energy E are

$$\psi_{n_x n_y n_z}(x, y, z) = \sqrt{8/(abc)} \sin(n_x \pi x/a) \sin(n_y \pi y/b) \sin(n_z \pi z/c) \qquad (3.36)$$

and

$$E_{n_x n_y n_z} = \frac{n_x^2 h^2}{8ma^2} + \frac{n_y^2 h^2}{8mb^2} + \frac{n_z^2 h^2}{8mc^2} \qquad (3.37)$$

where $n_x = 1, 2, 3, ...$; $n_y = 1, 2, 3, ...$; $n_z = 1, 2, 3, ...$

For the three–dimensional problem, three quantum numbers are necessary to describe the states of the system, $(n_x n_y n_z)$. The same energy can be obtained by several different sets of quantum numbers; such a state is said to be degenerate, otherwise nondegenerate, or to have a degeneracy $g = 1$.

For example, for a particle in a three–dimensional cubic potential box with $c = b = a$ the degeneracies of the lowest energy states are shown in Table 3.1 .

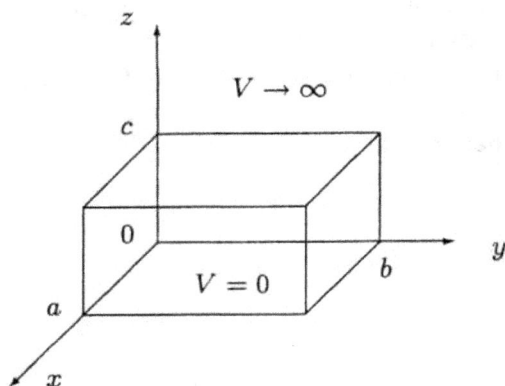

Fig. 3.2. A three–dimensional box.

Table 3.1: Energy and degeneracy of a particle in a cubic box.

State $n_x n_y n_z$	Energy $\times[h^2/(8ma^2)]$	Degeneracy g	Explanation
1 1 1	3	1	nondegenerate (ground state)
2 1 1	6	3	
1 2 1	6	3	3–fold degenerate
1 1 2	6	3	
1 2 2	9	3	
2 1 2	9	3	3–fold degenerate
2 2 1	9	3	
3 1 1	11	3	
1 3 1	11	3	3–fold degenerate
1 1 3	11	3	
2 2 2	12	1	nondegenerate

3.6   The wave equation for a particle with potential

The Schrödinger equation for a particle of mass m in the presence of a potential $V(x, y, z)$ may be written as

$$-\frac{\hbar^2}{2m}\left(\frac{\partial^2}{\partial x^2} + \frac{\partial^2}{\partial y^2} + \frac{\partial^2}{\partial z^2}\right)\psi(x, y, z) + V(x, y, z)\psi(x, y, z) = E\psi(x, y, z)$$

$$(3.38)$$

The first term on the left–hand side of Eq.(3.38) is associated with the kinetic energy T of the particle, and the second term with the potential energy V.

The average values of T and V are

$$<T> = \int\int\int \psi^*(x, y, z)T\psi(x, y, z)dxdydz \qquad (3.39)$$

$$<V> = \int\int\int \psi^*(x, y, z)V\psi(x, y, z)dxdydz \qquad (3.40)$$

Although the total energy of a system in a stationary state is well defined, in quantum mechanics only average values can be determined for its component parts, the kinetic energy and the potential energy.

3.7   Simple harmonic oscillator

Consider a particle of mass m is subject to a force F of the form (in one–dimension)

$$-\frac{dV}{dx} = F = -kx \qquad (Hooke's\ law\ force) \qquad (3.41)$$

where k is the force constant. For the minimum value of potential at $x = 0$, the corresponding potential is

$$V = \frac{1}{2}kx^2 = \frac{1}{2}m\omega^2 x^2 \qquad (harmonic\ oscillator\ potential) \qquad (3.42)$$

The corresponding Schrödinger equation is

$$-\frac{\hbar^2}{2m}\frac{d^2\psi(x)}{dx^2} + \frac{1}{2}kx^2\psi(x) = E\psi(x) \qquad (3.43)$$

The possible solution for this differential equation could be a simple Gaussian function

$$\psi(x) = Ae^{-\alpha^2 x^2/2} \;\; ; \;\; \alpha^2 = \frac{\sqrt{mk}}{\hbar} \;\; , \;\; A = \left(\frac{\alpha^2}{\pi}\right)^{1/4} \tag{3.44}$$

The corresponding energy is obtained as

$$E = \frac{1}{2}h\nu_0 = \frac{1}{2}\hbar\omega \tag{3.45}$$

This energy, Eq.(3.45), and the wave function, Eq.(3.44), correspond to the ground state. The general expressions are

$$E_n = (n + \frac{1}{2})h\nu_0 = (n + \frac{1}{2})\hbar\omega \;\; , \;\; n = 0, 1, 2, ... \tag{3.46}$$

and

$$\psi_n(x) = \left(\frac{\alpha}{\sqrt{\pi}} 2^n n!\right)^{1/2} H_n(\alpha x) e^{-\alpha^2 x^2/2} \tag{3.47}$$

where $H_n(\alpha x)$ are Hermite polynomials. The simple harmonic oscillator wave functions may also be expressed in the following form:

$$\psi_n(x) = \left(\frac{m\omega}{\hbar\pi}\right)^{1/4} \left(\frac{2^{-n}}{n!}\right)^{1/2} H_n(\xi) e^{-\xi^2/2} \;\; ; \;\; \xi = \left(\frac{m\omega}{\hbar}\right)^{1/2} x \tag{3.48}$$

The allowed energies are equally spaced, the constant interval being $h\nu_0$.

We note that the lowest state has the energy $\hbar\omega/2$. On the other hand the lowest energy of a classical harmonic oscillator is zero. The finite value $\hbar\omega/2$ of the ground state energy level, which is called the **zero–point energy**, is therefore a purely quantum mechanical effect, and is directly related to the uncertainty principle. The eigenvalues of the linear harmonic oscillator are non–degenerate, since for each eigenvalue there exists only one eigenfunction.

The Hermit polynomials $H_n(\xi)$ are defined as

$$H_n(\xi) = (-1)^n e^{\xi^2} \frac{d^n e^{\xi^2}}{d\xi^n} \tag{3.49}$$

these polynomials satisfy the recurrrence relation

$$H_{n+1}(\xi) - 2\xi H_n(\xi) + 2n H_{n-1}(\xi) = 0 \qquad (3.50)$$

The first few Hermite polynomials, obtained from Eq.(3.49), are

$$H_0(\xi) = 1 \qquad (3.51)$$

$$H_1(\xi) = 2\xi$$

$$H_2(\xi) = 4\xi^2 - 2$$

$$H_3(\xi) = 8\xi^3 - 12\xi$$

$$H_4(\xi) = 16\xi^4 - 48\xi^2 + 12$$

$$H_5(\xi) = 32\xi^5 - 160\xi^3 + 120\xi$$

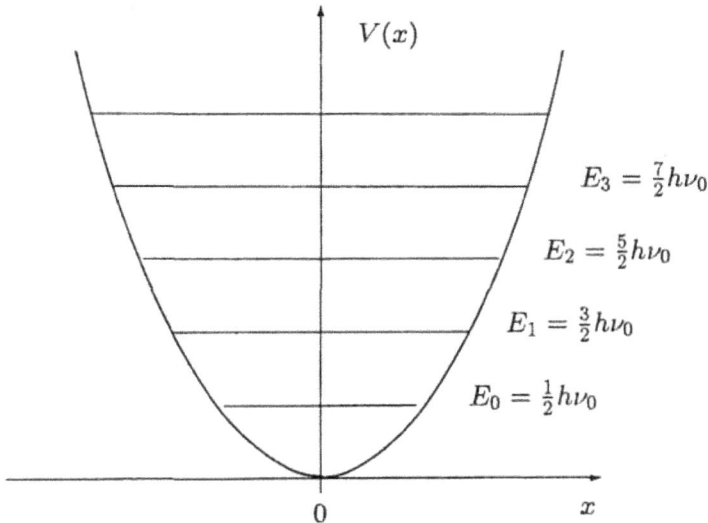

Fig. 3.3. Simple harmonic oscillator potential.

Table 3.2: Summary: Energies and wave functions of various systems.

System	Energy	Wave function
Rutherford atom	$E = -mZ^2e^4/(2p_\phi^2)$ continuous	no wave function
Bohr atom	$E_n = -mZ^2e^4/(2\hbar^2n^2)$ $n = 1, 2, 3, ...$ quantized	no wave function
Free particle	$E = \hbar^2k^2/(2m)$ continuous	$\psi(x) = A\cos(kx) + B\sin(kx)$
Particle in 1D–box	$E_n = h^2n^2/(8mL^2)$ $n = 1, 2, 3, ...$ quantized	$\psi_n(x) = \sqrt{2/L}\sin(n\pi x/L)$
Harmonic oscillator $V = kx^2/2$	$E_n = (n + 1/2)h\nu_0$ $n = 0, 1, 2, ...$ quantized	$\psi_n = A_nH_n(\alpha x)e^{-\alpha^2x^2/2}$ $\alpha^2 = \sqrt{mk}/\hbar$ $A_n = (\alpha/(\sqrt{\pi}2^nn!))^{1/2}$

3.8   Worked examples

Example - 3.1 :

Assume that an electron is in a cubic box with side 1Å. The potential outside the box is infinite, and it is zero inside the box. Calculate the possible lowest energy of the electron.

Solution :

$E_{n_x n_y n_z} = \frac{h^2}{8m}\left(\frac{n_x^2}{a^2} + \frac{n_y^2}{b^2} + \frac{n_z^2}{c^2}\right)$ lowest value is for $n_x = n_y = n_z = 1$,

for cubic box $a = b = c$ $E_{111} = \frac{3h^2}{8ma^2} = \frac{3}{8}\frac{4\pi^2\hbar^2}{ma^2} = \frac{3\pi^2\hbar^2}{2ma^2}$

in a.u. $\hbar = 1$, $m = 1$, $a = 1/0.529$, 1 a.u. $\cong 27.21\ eV$

$E_{111} = \frac{3\pi^2}{2(1/0.529)^2}$ a.u. $= 1.5 \times \pi^2 \times (0.529)^2$ a.u. $\cong 4.143$ a.u. $= 112.73\ eV$.

Example - 3.2 :

A system described by the Hamiltonian

$$H = -\frac{\hbar^2}{2m}\nabla^2 + \frac{1}{2}m(\omega_1^2 x^2 + \omega_2^2 y^2 + \omega_3^2 z^2)$$

is called an *anisotropic harmonic oscillator*. Determine the possible energies of this system, and, for the isotropic case ($\omega_1 = \omega_2 = \omega_3$), calculate the degeneracy of the level E_n.

Solution :

Since $V(x, y, z) = V_1(x) + V_2(y) + V_3(z)$, this problem reduces to the problem of three independent harmonic oscillators of frequencies ω_1, ω_2, ω_3, along the axes x, y, z, respectively.

Therefore,
$$E_{n_1 n_2 n_3} = \hbar\omega_1(n_1 + 1/2) + \hbar\omega_2(n_2 + 1/2) + \hbar\omega_3(n_3 + 1/2)$$

$$\psi_{n_1 n_2 n_3} = \left(\frac{m^3 \omega_1 \omega_2 \omega_3}{\hbar^3 \pi^3}\right)^{\frac{1}{4}} \left(\frac{2^{-(n_1+n_2+n_3)}}{n_1! n_2! n_3!}\right)^{\frac{1}{2}} H_{n_1}(\xi_1) H_{n_2}(\xi_2) H_{n_3}(\xi_3)$$

$$\times e^{-\frac{1}{2}(\xi_1^2 + \xi_2^2 + \xi_3^2)}$$

where $\xi_1 = (m\omega_1/\hbar)^{1/2}x$, $\xi_2 = (m\omega_2/\hbar)^{1/2}y$, $\xi_3 = (m\omega_3/\hbar)^{1/2}z$ and $n_1, n_2, n_3 = 0, 1, 2, ...$ If the ratios of the frequencies are irrational, the energy levels are non–degenerate, otherwise they may be degenerate. The ground state E_{000} is always non–degenerate. For the isotropic harmonic oscillator, $E_n = \hbar\omega(n + 3/2)$, where $n = n_1 + n_2 + n_3$. In this case all the energy levels with the exception of E_0 are degenerate. To calculate the degeneracy of the level of energy E_n, consider for the moment a particular value of the quantum number n_1. n_2 can then have any of the values $0, 1, ..., n - n_1$, and the sum $n = n_1 + n_2 + n_3$ for given n and n_1 can be obtained in $n - n_1 - 1$ ways. Since $n_1 = 0, 1, 2, ..., n$, the degeneracy of E_n will be $\sum_{n_1=0}^{n}(n - n_1 + 1) = \frac{1}{2}(n + 1)(n + 2)$.

Example - 3.3 :

Determine the current density $\mathbf{j}$ for the plane wave

$$\psi(x,t) = Ae^{\frac{i}{\hbar}[\sqrt{2mE}x - Et]}$$

Solution :

In one–dimension $j = \frac{i\hbar}{2m}\left(\psi\frac{d\psi^*}{dx} - \psi^*\frac{d\psi}{dx}\right)$. Then

$$j = \frac{i\hbar}{2m}\left(\psi[\frac{-i}{\hbar}\sqrt{2mE}]\psi^* - \psi^*[\frac{i}{\hbar}\sqrt{2mE}]\psi\right)$$

$$= \frac{\sqrt{2mE}}{m}\psi\psi^* = \frac{p_x}{m}|\psi|^2 = v_x|\psi|^2$$

This expression gives the number of particles passing through a unit area of a plane perpendicular to the x–axis in one second.

Example - 3.4 :

An electron is trapped in a well defined by the potential energy function $V(x) = -e^2/[(4\pi\varepsilon_0)x]$ for $x > 0$; $V(x) = \infty$ for $x \leq 0$. Write the wave equation for this system in terms of the variable z instead of x by using the definitions: $x = \frac{1}{2}\alpha a_0 z$ where $\alpha^{-2} = -(2mEa_0^2)/\hbar^2$ and $a_0 = \hbar^2(4\pi\varepsilon_0)/(me^2)$.

Solution :

The Schrödinger equation is

$$Hu = \left(-\frac{\hbar^2}{2m}\frac{d^2}{dx^2} - \frac{e^2}{(4\pi\varepsilon_0)x}\right)u = Eu \quad , \quad x > 0$$

The suggested transformation gives in a straightforward manner

$$\frac{d^2u}{dz^2} - \frac{1}{4}u + \frac{\alpha}{z}u = 0$$

This differential equation may be identified as Kummer's equation which has as its general solution the confluent hypergeometric Wittaker Functions.

Chapter 4

Perturbation Theory and Radiative Transitions

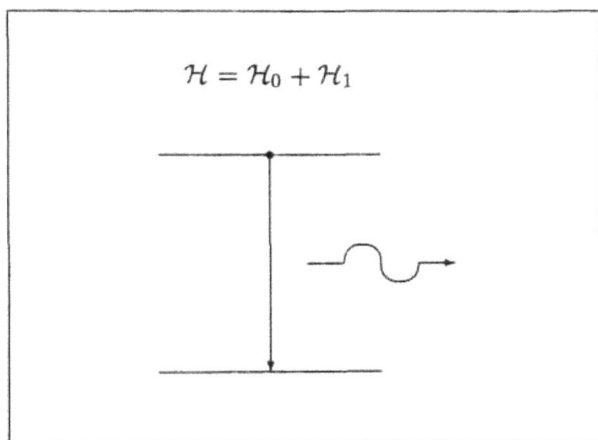

$$\mathcal{H} = \mathcal{H}_0 + \mathcal{H}_1$$

4.1   Time–independent perturbation theory

Perturbation theory is one of the important and frequently used approximation methods. There are two types of perturbation theory, time–independent perturbation theory, used to find the stationary states of a system, and time–dependent perturbation theory, used to calculate certain quantities called transition probabilities.

Time–independent perturbation theory is an approximation method which is applied to time–independent Schrödinger equation. The formalism of time–independent perturbation theory is a little bit different for non–degenerate and degenerate states.

Non–degenerate perturbation theory:

We assume that we have found the eigenvalues and the orthonormal complete set of eigenfunctions for a Hamiltonian H_0,

$$H_0 \phi_j = E_j^{(0)} \phi_j \tag{4.1}$$

and we ask for the eigensolutions for the Hamiltonian

$$H = H_0 + \lambda H_1 \tag{4.2}$$

that is, for the solutions of

$$(H_0 + \lambda H_1)\psi_j = E_j \psi_j \tag{4.3}$$

Here H_1 is assumed to be small and called as the perturbation. The desired quantities, ψ_j and E_j, may be expressed as power series in λ, which is a constant. We will assume that as $\lambda \to 0$, $E_j \to E_j^{(0)}$ and $\psi_j \to \phi_j^{(0)}$. Since the ϕ_j form an orthonormal complete set, we may expand ψ_j in a series involving all the ϕ_j. We write

$$\psi_j = N(\lambda) \left[\phi_j + \sum_{i \neq j} C_{ji}(\lambda)\phi_i \right] \tag{4.4}$$

The factor $N(\lambda)$ is for the normalization of ψ_j; we have $N(0) = 1$, $C_{ji}(0) = 0$. More generally we have

$$C_{ji} = \lambda C_{ji}^{(1)} + \lambda^2 C_{ji}^{(2)} + \cdots \quad ; \quad E_j = E_j^{(0)} + \lambda E_j^{(1)} + \lambda^2 E_j^{(2)} + \cdots \tag{4.5}$$

After substituting Eqs.(4.4) and (4.5) in (4.3) and identifying powers of λ yields a series of equations. The first three of these equations are:

$$H_0 \phi_j = E_j^{(0)} \phi_j \tag{4.6}$$

$$H_0 \sum_{i \neq j} C_{ji}^{(1)} \phi_i + H_1 \phi_j = E_j^{(0)} \sum_{i \neq j} C_{ji}^{(1)} \phi_i + E_j^{(1)} \phi_j \tag{4.7}$$

$$H_0 \sum_{i \neq j} C_{ji}^{(2)} \phi_i + H_1 \sum_{i \neq j} C_{ji}^{(1)} \phi_i = E_j^{(0)} \sum_{i \neq j} C_{ji}^{(2)} \phi_i + E_j^{(1)} \sum_{i \neq j} C_{ji}^{(1)} \phi_i + E_j^{(2)} \phi_j \tag{4.8}$$

Let us consider the equation for λ^1, Eq.(4.7): If we take the scalar product of Eq.(4.7) with ϕ_j, we obtain

$$E_j^{(1)} = < \phi_j | H_1 | \phi_j > \tag{4.9}$$

This expression states that the first order energy shift for a given state is just the expectation value of the perturbing potential in that state. If we take the scalar product of Eq.(4.7) with ϕ_k for $k \neq j$, we obtain

$$C_{ji}^{(1)} = \frac{< \phi_i | H_1 | \phi_j >}{E_j^{(0)} - E_i^{(0)}} \quad , \quad (j \neq i) \tag{4.10}$$

The numerator of $C_{ji}^{(1)}$ is the matrix element of H_1 in the basis of states in which H_0 is diagonal.

Let us now consider the equation for λ^2, Eq.(4.8): Taking the scalar product with ϕ_j yields

$$E_j^{(2)} = \sum_{i \neq j} < \phi_j | H_1 | \phi_i > C_{ji}^{(1)} \tag{4.11}$$

$$= \sum_{i \neq j} \frac{< \phi_j | H_1 | \phi_i > < \phi_i | H_1 | \phi_j >}{E_j^{(0)} - E_i^{(0)}} = \sum_{i \neq j} \frac{| < \phi_i | H_1 | \phi_j > |^2}{E_j^{(0)} - E_i^{(0)}}$$

This expression states that the second order energy shift is the sum of terms, whose strength is given by the square of the matrix element connecting the given state ϕ_j to all other states by the perturbation potential, weighted by the reciprocal of the energy difference between the states.

An expression for $C_{ji}^{(2)}$ may be obtained from the scalar product of Eq.(4.8) with ϕ_k, for $k \neq j$; the result becomes

$$C_{ji}^{(2)} = \sum_{k\neq j} \frac{< \phi_j|H_1|\phi_k >< \phi_k|H_1|\phi_j >}{(E_j^{(0)} - E_i^{(0)})(E_j^{(0)} - E_k^{(0)})} - E_j^{(1)} \frac{< \phi_i|H_1|\phi_j >}{(E_j^{(0)} - E_i^{(0)})^2} \quad (4.12)$$

The normalization factor $N(\lambda)$ can be determined from

$$< \psi_j|\psi_j >= N^2(\lambda) \left[1 + \lambda^2 \sum_{i\neq j} |C_{ji}^{(1)}|^2 + \cdots \right] = 1 \quad (4.13)$$

It is therefore 1 to first order in λ. Hence, to first order in λ, we may write

$$\psi_j = \phi_j + \lambda \sum_{i\neq j} \frac{< \phi_i|H_1|\phi_j >}{E_j^{(0)} - E_i^{(0)}} \phi_i \quad (4.14)$$

To second order in λ, we may write

$$\psi_j = \left[1 + \lambda^2 \sum_{i\neq j} |C_{ji}^{(1)}|^2\right]^{-1/2} \left[\phi_j + \lambda \sum_{i\neq j} C_{ji}^{(1)}\phi_i + \lambda^2 \sum_{i\neq j} C_{ji}^{(2)}\phi_i\right] \quad (4.15)$$

Degenerate perturbation theory:

In the degenerate case, instead of a unique ϕ_j, there is a finite set of $\phi_j^{(k)}$, all of which have the same energy $E_j^{(0)}$. We choose the set of $\phi_j^{(k)}$ such that $< \phi_i^{(\ell)}|\phi_j^{(k)} >= \delta_{ij}\delta_{\ell k}$. The wave function in this case may be expanded as follows:

$$\psi_j = N(\lambda) \left[\sum_k \alpha_k\phi_j^{(k)} + \lambda \sum_{i\neq j} C_{ji}^{(1)} \sum_k \beta_k\phi_i^{(k)} + \cdots \right] \quad (4.16)$$

The coefficients α_k , β_k , $\cdots$ will have to be determined. When this wave function is used in the Schrödinger equation, Eq.(4.3), we get, to first order in λ:

$$H_0 \sum_{i\neq j} C_{ji}^{(1)} \sum_k \beta_k\phi_i^{(k)} + H_1 \sum_k \alpha_k\phi_j^{(k)} = \quad (4.17)$$

$$E_j^{(1)} \sum_k \alpha_k \phi_j^{(k)} + E_j^{(0)} \sum_{i \neq j} C_{ji}^{(1)} \sum_k \beta_k \phi_i^{(k)}$$

Taking the scalar product with $\phi_j^{(\ell)}$ gives the first order shift equation

$$\sum_k \alpha_k < \phi_j^{(\ell)} |H_1| \phi_j^{(k)} > = E_j^{(1)} \alpha_\ell \tag{4.18}$$

This is a finite–dimensional eigenvalue problem. For example, if there is an m–fold degeneracy, and if we use the notation $< \phi_j^{(n)} |H_1| \phi_j^{(m)} > = h_{nm}$, this equation takes the matrix form

$$\begin{pmatrix} h_{11} & \cdots & h_{1m} \\ \vdots & & \vdots \\ h_{m1} & \cdots & h_{mm} \end{pmatrix} \begin{pmatrix} \alpha_1 \\ \vdots \\ \alpha_m \end{pmatrix} = E_j^{(1)} \begin{pmatrix} \alpha_1 \\ \vdots \\ \alpha_m \end{pmatrix} \tag{4.19}$$

Both the eigenvalues and the α_k, can be determined from this equation, if we add the condition that $\sum_{k=1}^m |\alpha_k|^2 = 1$.

4.2 Time–dependent perturbation theory

We now discuss the behaviour of a system under the influence of a perturbation that varies with time:

$$H = H_0 + H_1(t) \tag{4.20}$$

We suppose that, as before, H_0 does not vary with t. We again use the stationary– state wave functions of H_0 to make an expansion of the wave function $\Psi(\mathbf{r}, t)$ of the system. Since we are interested in the time dependence of Ψ, we must now include in the stationary–state wave functions of H_0 their time variation:

$$\Phi_j(\mathbf{r}, t) = \phi_j(\mathbf{r}, t)e^{-iE_j t/\hbar} \tag{4.21}$$

where the $\phi_j(\mathbf{r})$ are the same eigen functions as described in time–independent perturbation theory. Because H varies with t, it does not have stationary–state wave functions; we must solve the time–dependent Schrödinger equation

$$H\Psi_j(\mathbf{r}, t) = i\hbar \frac{\partial \Psi_j(\mathbf{r}, t)}{\partial t} \tag{4.22}$$

Suppose that as $t \to -\infty$, $H_1(t) \to 0$, so that initially H coincides with H_0. Thus initially the system may be in a stationary–state of H_0. Suppose that the system initially in the stationary–state i, so the wave function Ψ_i coincides with Φ_i at $t = -\infty$. When the perturbation $H_1(t)$ begins to switch on, ψ_i will start to become different from Φ_i. We use the expansion

$$\Psi_i(\mathbf{r}, t) = \sum_j a_{ij}(t) \Phi_j(\mathbf{r}, t) \tag{4.23}$$

The coefficients a_{ij} are functions of t, because the Φ_j are not stationary–state wave functions for the perturbed system. At $t = -\infty$, $a_{ii} = 1$, and all the other a_{ij} are zero. Inserting this expansion into the time–dependent Schrödinger equation, and using the equation

$$H_0 \Phi_j(\mathbf{r}, t) = i\hbar \frac{\partial \Phi_j}{\partial t} \tag{4.24}$$

which defines Φ_j, we find that

$$i\hbar \sum_j \frac{\partial a_{ij}}{\partial t} \Phi_j = \sum_j a_{ij} H_1 \Phi_j \quad . \tag{4.25}$$

After pre–multiplying this equation by Φ_k^*, and integrating over all space, we obtain, with the help of the orthonormality conditions

$$< \Phi_k | \Phi_j > = \delta_{kj} \tag{4.26}$$

$$i\hbar \frac{\partial a_{ik}}{\partial t} = \sum_j a_{ij} < \Phi_k | H_1 | \Phi_j > \tag{4.27}$$

So far no approximation is made. The last equation is exactly equivalent to the time–dependent Schrödinger equation. We now suppose that H_1 is small. This means that we expect the coefficient functions $a_{ij}(t)$ to depart rather little from their original values, at least if the time t is not too large. We will suppose that the value of t is such that $a_{ii}(t)$ is still close to 1, and that $a_{ij}(t)$ for $i \neq j$ is still small. We then take care of the terms in Eq.(4.27) that are first order in small quantities, which means that on the right–hand side we keep only the term in the sum for which $j = i$ and in it we replace a_{ii} by 1:

$$i\hbar \frac{\partial a_{ik}(t)}{\partial t} = < \Phi_k | H_1 | \Phi_i > = < \phi_k | H_1 | \phi_i > e^{i(E_k - E_i)t/\hbar} \tag{4.28}$$

Integrating with respect to t, we have

$$a_{ik}(t) = \delta_{ik} - \frac{i}{\hbar} \int_{-\infty}^{t} dt' < \phi_k |H_1(t')|\phi_i > e^{i(E_k - E_i)t'/\hbar} \qquad (4.29)$$

where the δ_{ik} is the integration constant that takes account of the initial conditions at $t = -\infty$.

4.3 Radiative transitions

When a system changes its energy as the result of the emission or the absorption of a photon, it is said to undergo radiative transition. There are three kinds of radiative transition. In the presence of an electromagnetic field the system can absorb a photon, so that its energy is raised to a higher level. If the system is initially in a state other than the ground state, it can emit a photon and thus lose energy. This can happen without the presence of any external electromagnetic field, in which case a spontaneous emission is said to occur. On the other hand, if the excited system is placed in an electromagnetic field that varies in time with the appropriate frequency, the probability that it emits a photon can be greatly increased. In this case the emission process is known as stimulated emission. Stimulated emission is the basis of maser and laser action.

In practice, the intensity of the external electromagnetic field is usually large compared with that of the field associated with the single photon that is being absorbed or emitted. In such a case we can consider a semiclassical calculation, where the external field is not quantized. We are interested in the interaction of the electromagnetic field with a charged particle, for example with an electron in an atom. This means that we must incorporate the effect of the field into the Hamiltonian for the quantum system that we are studying.

For a particle moving in a potential $V(\mathbf{r})$, the Hamiltonian is

$$H_0 = \frac{p^2}{2m} + V(\mathbf{r}) \ . \qquad (4.30)$$

Classical Hamilton's equations for the electron are

$$\dot{\mathbf{r}} = \frac{\mathbf{p}}{m} \quad , \quad \dot{\mathbf{p}} = -\nabla V = m\ddot{\mathbf{r}} \qquad (4.31)$$

The external electromagnetic field may be described by a scalar potential $\phi(\mathbf{r}, t)$ and a vector potential $\mathbf{A}(\mathbf{r}, t)$, so that its electric and magnetic vectors are given by

$$\vec{\mathcal{E}} = -\nabla\phi - \mathbf{A} \quad , \quad \mathbf{B} = \nabla \times \mathbf{A} \tag{4.32}$$

The Hamiltonian then changes from H_0 to

$$H = V + (\mathbf{p} + e\mathbf{A})^2/2m - e\phi \tag{4.33}$$

Hamilton's equations, Eq.(4.31), together with Eq.(4.32), now give

$$m\ddot{\mathbf{r}} = -\nabla V - e\vec{\mathcal{E}} - e\dot{\mathbf{r}} \times \mathbf{B} \tag{4.34}$$

The last two terms of this equation are just the Lorentz force of the electromagnetic field on the electron. When $\mathbf{p}$ is replaced by $-i\hbar\nabla$ and the gauge condition

$$\nabla \cdot \mathbf{A} = 0 \tag{4.35}$$

taken into account the Hamiltonian, Eq.(4.33), may be expressed as follows

$$H = H_0 + H_1 \quad , \quad H_1 = -i\hbar\frac{e}{m}\mathbf{A} \cdot \nabla - e\phi \tag{4.36}$$

In the absence of the electromagnetic field, the electron wave function satisfies the Schrödinger equation with Hamiltonian H_0. Let the stationary–state wave functions of H_0 be

$$\Phi_k = \phi_k e^{-iE_k t/\hbar} \tag{4.37}$$

corresponding to energies E_k.

In reality, only the ground state is a strictly stationary state, because the higher states can spontaneously emit radiation and so change their configuration. By taking all the levels E_k to be stationary states, we are making an approximation in which the possibility of spontaneous emission is neglected. Here we will consider only the stimulated emission (and absorption) induced by the external electromagnetic field. We will take this field to be the plane wave given by

$$\phi = 0 \quad , \quad \mathbf{A} = \mathbf{A}_0 \sin(\mathbf{k} \cdot \mathbf{r} - \omega t) \tag{4.38}$$

The corresponding electric and magnetic vectors are

$$\vec{\mathcal{E}} = \omega A_0 \cos(\mathbf{k} \cdot \mathbf{r} - \omega t) \quad , \quad \mathbf{B} = \mathbf{k} \times \mathbf{A}_0 \cos(\mathbf{k} \cdot \mathbf{r} - \omega t) \qquad (4.39)$$

In order to satisfy Maxwell's equations, which are given by

$$\nabla \cdot \vec{\mathcal{E}} = 0 \quad , \quad \nabla \cdot \mathbf{B} = 0 \quad , \quad \nabla \times \vec{\mathcal{E}} = -\frac{\partial \mathbf{B}}{\partial t} \quad , \quad \nabla \times \mathbf{B} = \frac{1}{c^2}\frac{\partial \vec{\mathcal{E}}}{\partial t} \qquad (4.40)$$

we impose the conditions:

$$\omega = ck \quad , \quad \mathbf{k} \cdot \mathbf{A}_0 = 0 \qquad (4.41)$$

We will choose the directions of the coordinate axes such that the direction of propagation, parallel to the wave vector $\mathbf{k}$, is in the z–direction, and such that $\mathbf{A}_0$ is in the x–direction. Then,

$$H_1 = -i\hbar \frac{e A_0}{m} \left(\frac{\partial}{\partial x} \right) \sin(kz - \omega t) \qquad (4.42)$$

$$= -\frac{\hbar e A_0}{2m} \left(\frac{\partial}{\partial x} \right) \left[e^{i(kz - \omega t)} - e^{-i(kz - \omega t)} \right]$$

Transition Probability:

For weak fields we can use the time–dependent perturbation theory. We suppose that the field is weak. Imagine that the field is switched on at time $t = 0$ and is switched off again at some time T. We suppose that the electron was initially in the eigenstate Φ_i of H_0, and calculate the probability that a measurement, after the field has been switched off, will find that a transition has occured to some other eigenstate of H_0. This transition probability is the square modulus of the coefficient function $a_{if}(t)$ described in time–dependent perturbation theory, namely $|a_{if}(t)|^2$. Remembering that, Eq.(4.29):

$$a_{if}(t) = \delta_{if} - \frac{i}{\hbar} \int_{-\infty}^{t} <\phi_f|H_1(t')|\phi_i> e^{i(E_f - E_i)t'/\hbar}$$

we may rewrite this coefficient after substituting $H_1(t)$

$$a_{if} = -\frac{i}{\hbar} \int_0^T dt \left[u_{fi}^{(+)} e^{i(\omega_{fi} - \omega)t} - u_{fi}^{(-)} e^{i(\omega_{fi} + \omega)t} \right] \qquad (4.43)$$

where

$$\omega_{fi} = (E_f - E_i)/\hbar \tag{4.44}$$

and

$$u_{fi}^{(\mp)} = -\frac{ieA_0}{2m} < \phi_f | \frac{\partial}{\partial x} e^{\mp ikz} | \phi_i > \tag{4.45}$$

If $E_f > E_i$, the transition must have occured through the absorption of energy from the electromagnetic field, and the electron has absorbed a single photon from the field. If $E_f < E_i$, the electron has emitted a single photon. In either case, the transition probability is proportional to $|a_{if}|^2$.

After integration in a_{if} we get

$$a_{if} = \frac{1}{\hbar} \left[\frac{1 - e^{i(\omega_{fi} - \omega)T}}{\omega_{fi} - \omega} u_{fi}^{(+)} - \frac{1 - e^{i(\omega_{fi} + \omega)T}}{\omega_{fi} + \omega} u_{fi}^{(-)} \right] \tag{4.46}$$

The angular frequency ω of the external field can be tuned to give a resonance in the transition probability. This condition is obtained when $\omega_{fi}/\omega \ll 1$. In the case of absorption, where $E_f > E_i$, so that $\omega_{fi} > 0$, the first term in a_{if} becomes dominant and the transition probability is

$$|a_{if}|^2 = \frac{4}{\hbar^2} |u_{fi}^{(+)}|^2 \cdot \frac{\sin^2[(\omega_{fi} - \omega)T/2]}{(\omega_{fi} - \omega)^2} \tag{4.47}$$

In the case of emission, where $\omega_{fi} < 0$, the second term in a_{if} leads to an analogous expression.

Electric Dipole Transitions:

In order to evaluate the transition probability $|a_{if}|^2$, it is necessary to calculate $u_{fi}^{(\mp)}$, where

$$u_{fi}^{(\mp)} \propto < \phi_f | \hat{p}_x e^{\mp ikz} | \phi_i > \tag{4.48}$$

Since ϕ_i and ϕ_f are electron wave functions corresponding to atomic levels, they are small unless $kr \ll 1$. Here it is a good approximation to replace the exponential in $u_{fi}^{(\mp)}$ by unity:

$$u_{fi}^{(\mp)} \cong -i\frac{eA_0}{2m} < \phi_f | \hat{p}_x | \phi_i > \tag{4.49}$$

Using the commutation relation

$$\hat{p}_x = -\frac{i}{\hbar}m[x, H_0] \tag{4.50}$$

and remembering that ϕ_i and ϕ_f are eigenfunctions of H_0, we obtain

$$u_{fi}^{(\mp)} = \frac{A_0(E_i - E_f)}{2\hbar} < \phi_f| - ex|\phi_i > \tag{4.51}$$

Now, $-ex$ is the x–component of the electric dipole moment of the atom. For this reason, the transitions corresponding to the approximation we have made, that $e^{i\mp kz} \approx 1$, are known as electric dipole transitions. If we had not replaced the exponential $e^{i\mp kz}$ by unity, we should have obtained additional terms corresponding to electric quadrupole and higher multipole transitions, and also terms corresponding to magnetic dipole and multipole moments.

The transition probabilities are greatest for the electric dipole transitions. But for many pairs of levels E_i and E_f the quantity $< \phi_f|-ex|\phi_i >$ vanishes. This leads to certain selection rules: **radiative transitions occur only between certain pairs of atomic levels**. We will discuss selection rules in detail in the next chapter.

Spectroscopic Transitions:

Dipole transitions: $\mu = er$, electric dipole moment, transition dipole moment, for absorption

$$\mu_{nm} =< \psi_n|\mu|\psi_m > \tag{4.52}$$

for the reverse case, emission,

$$\mu_{mn} =< \psi_m|\mu|\psi_n > \tag{4.53}$$

$(\mu_{mn} = \mu_{nm})$.

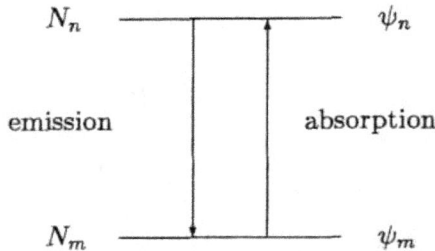

Fig. 4.1. Spectroscopic transitions.

Absorption and emission that take place in the presence of an external field are called the **induced absorption** or **emission**. The system in state n may spontaneously lose an amount of energy E given by

$$E = E_n - E_m = h\nu \tag{4.54}$$

We now derive a relationship, according to Einstein, for the transition probabilities of induced absorption and emission and spontaneous emission.

According to the Maxwell–Boltzmann distribution law, the ratio of the population in state n and m, N_n/N_m, is given by (in a thermal equilibrium at a given temperature T)

$$\frac{N_n}{N_m} = e^{-(E_n - E_m)/kT} \tag{4.55}$$

k is the Boltzmann constant. According to the Maxwell–Boltzmann law the number of excited atoms will increase with increasing temperature.

We define the transition probability per unit time:
For the induced absorption: B_{mn}
For the induced emission   : B_{nm}
For the spontaneous emission: A_{nm}
($B_{mn} = B_{nm}$). The unit of these quantities is s^{-1}.

The number of induced emission per second is proportional to $N_n B_{nm}$ and is given by $N_n B_{nm} \rho(\nu)$, where $\rho(\nu)$ is the density of the radiation of frequency ν. The radiation density, $\rho(\nu)$, is defined as energy per unit volume.

The number of induced absorption per second is given by $N_m B_{mn}\rho(\nu)$. The number of transitions per second for a spontaneous emission is $N_n A_{nm}$.

At steady state (at thermodynamic equilibrium), the rate of absorption and emission are equal,

$$N_m B_{mn}\rho(\nu) = N_n B_{nm}\rho(\nu) + N_n A_{mn} \tag{4.56}$$

Since $B_{mn} = B_{nm}$, we write

$$\rho(\nu) = \frac{N_n A_{nm}}{B_{nm}(N_m - N_n)} = \frac{A_{nm}}{B_{nm}} \frac{1}{e^{(E_m - E_n)/kT} - 1} \tag{4.57}$$

According to Planck's radiation law (black–body radiation)

$$\rho(\nu) = \frac{8\pi h\nu^3}{c^3} \frac{1}{e^{h\nu/kT} - 1} \tag{4.58}$$

$$\frac{8\pi h\nu^3}{c^3} \frac{1}{e^{h\nu/kT} - 1} = \frac{A_{nm}}{B_{nm}} \frac{1}{e^{(E_m - E_n)/kT} - 1} \tag{4.59}$$

Since $\nu = \nu_{nm}$, we have

$$\frac{A_{nm}}{B_{nm}} = \frac{8\pi h\nu_{nm}^3}{c^3} \tag{4.60}$$

The intensity of the spontaneous emission line is

$$I_{em.} = h\nu A_{nm} N_n \tag{4.61}$$

In general, therefore, the intensity of the absorption lines is proportional to the population in the lower state.

When a sample is irradiated there will also be induced and spontaneous emission and we might expect that this competing effect will diminish the intensity of the absorption line. However, since only the induced emission is coherent with the incident radiation, that is, it travels in the same direction of the radiation field, the net absorption intensity is given by

$$I_{ab.} = h\nu[N_m B_{mn}\rho(\nu) - N_n B_{nm}\rho(\nu)] = B_{mn}\rho(\nu)(N_m - N_n)h\nu \quad (4.62)$$

Thus the intensity for a given absorption depends on the difference in the population in states m and n if $\rho(\nu)$ is kept constant.

4.4   Line width and line broadening

Because of the spontaneous emission, any transition line in reality has a width which is determined by the finite lifetime of the upper state. This is most clearly shown by the Heisenberg uncertainty principle $\Delta E \Delta t \cong \hbar$. Thus we can no longer represent the upper level by a sharp line (which represents only one value of E); the resulting transition will have an uncertainty in energy ΔE, which is the **linewidth** (Γ). Such a broadening in the spectral lines are called as the **lifetime broadening**.

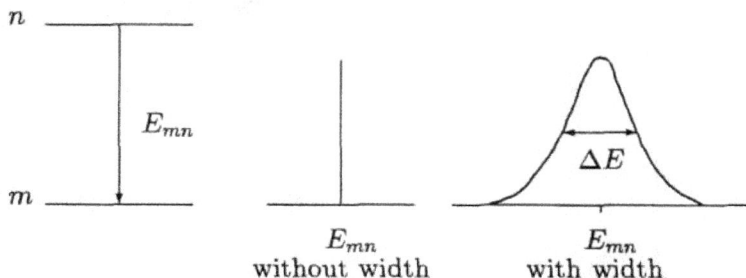

Fig. 4.2. Emission and line width.

The intensity profile of transition lines looks–like Lorentzian distribution. Lorentzian intensity distribution:

$$f(\omega) = \frac{\Gamma^2/4\hbar^2}{(\omega - \omega_{nm})^2 + \Gamma^2/4\hbar^2} \quad (4.63)$$

$$\Gamma = \frac{\hbar}{\tau_m} + \frac{\hbar}{\tau_n} \quad or \quad \Delta\nu_N = \frac{1}{2\pi}\left(\frac{1}{\tau_m} + \frac{1}{\tau_n}\right) \quad (4.64)$$

where τ is the half–life of the corresponding state. $\tau_m \rightarrow \infty$ if m is the ground state, then $\hbar/\tau_n$ is the **natural width** of the state n.

$$\Delta\nu_N = \frac{A_{nm}}{2\pi} \quad (4.65)$$

for a level making a transition to the ground state. $\Delta \nu_N$ is in the order of 10^{-5} nm.

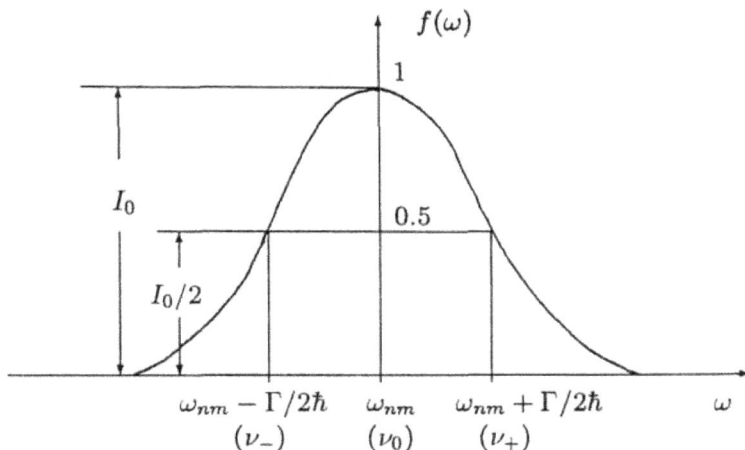

Fig. 4.3. Lorentzian intensity distribution.

In addition to the lifetime broadening, spectral lines can also be broadened by the Doppler effect. This arises when the species being measured has a velocity v relative to the observing instrument.

If the species is moving away from the instrument with velocity $-v$, the observed frequency of radiation ν' (by the species) is given by

$$\nu' = \nu(1 - v/c) \quad or \quad \frac{\nu - \nu'}{\nu} = \frac{\Delta \nu}{\nu} = \frac{v}{c} \quad \longrightarrow \quad \Delta \nu = \frac{v}{c}\nu \quad (4.66)$$

where ν is the frequency of the radiation field, and c the speed of light.

On the other hand, if the species is moving toward the instrument with velocity v, the observed frequency of radiation ν' (by the species) is given by

$$\nu' = \nu(1 + v/c) \quad or \quad \frac{\nu - \nu'}{\nu} = \frac{\Delta \nu}{\nu} = -\frac{v}{c} \quad \longrightarrow \quad \Delta \nu = -\frac{v}{c}\nu \quad (4.67)$$

Therefore, depending on the direction of motion, the observed frequency is shifted either toward the lower or higher frequency. The line broadening due to the Doppler effect may be calculated from the formula:

$$\Delta\nu_D = \frac{\nu}{c}\left(\frac{2RT}{M}\right)^{1/2} \tag{4.68}$$

$\Delta\nu_D$ is in the order of 10^{-4} nm.

One of the main causes of line broadening is due to the collision among species, in many cases this broadening can be related to the pressure of the gas. The pressure broadening ($\Delta\nu_P$) for the lines of transitions to ground state may be calculated from the formula:

$$\Delta\nu_P = \frac{1}{\pi}\sigma_L^2 N_2 \left[2\pi RT\left(\frac{1}{M_1}+\frac{1}{M_2}\right)\right]^{1/2} \tag{4.69}$$

where σ_L is collision cross–section, N_2 is the number of foring atoms and M are masses of atoms. ν_P is in the order of 10^{-5} nm.

Rate processes such as dissociation, rotation, electron and proton transfer reactions can also cause line broadening. Lines can also be broadened by the measuring instrument itself. We define **resolving power** of an instrument as a measure of its ability to distinguish lines that overlap. It is usually expressed as the ratio of the observed wavelength (or frequency) to the smallest difference between two wavelengths (or frequencies) that can be measured, that is $\lambda/\Delta\lambda$ (or $\nu/\Delta\nu$).

The spectral linewidth is mainly determined by the Doppler and pressure line broadening effects, and the total line broadening ($\Delta\nu_T$) may be written as

$$\Delta\nu_T = \frac{1}{2}\Delta\nu_P + \frac{1}{2}\left[(\Delta\nu_P)^2 + (\Delta\nu_D)^2\right]^{1/2} \tag{4.70}$$

4.5 Polarization of radiation

The electric $\mathbf{E}$ and magnetic $\mathbf{B}$ fields can be generated from scalar and vector potentials ϕ and $\mathbf{A}$ by

$$\mathbf{E}(\mathbf{r},t) = -\nabla\phi(\mathbf{r},t) - \frac{\partial\mathbf{A}(\mathbf{r},t)}{\partial t} \quad , \quad \mathbf{B}(\mathbf{r},t) = \nabla\times\mathbf{A}(\mathbf{r},t) \tag{4.71}$$

From Maxwell's equations, we write

$$\nabla \cdot \mathbf{A} = 0 \quad , \quad \nabla^2 \mathbf{A} - \frac{1}{c^2}\frac{\partial^2 \mathbf{A}}{\partial t^2} = 0 \tag{4.72}$$

A monochromatic plane wave solution of these equations represent a vector potential $\mathbf{A}$ as

$$\mathbf{A}(\omega; \mathbf{r}, t) = 2\mathbf{A}_0(\omega)\cos(\mathbf{k} \cdot \mathbf{r} - \omega t + \delta_\omega) \tag{4.73}$$

$$= \mathbf{A}_0(\omega)\left[e^{i(\mathbf{k}\cdot\mathbf{r} - \omega t + \delta_\omega)} + c.c.\right]$$

where $\mathbf{A}_0$ describes both the intensity and the polarization of the radiation, $\mathbf{k}$ is the propagation vector, ω is the angular frequency and δ_ω is the phase factor.

$$\mathbf{k} \cdot \mathbf{A}_0(\omega) = 0 \quad \longrightarrow \quad \mathbf{k} \perp \mathbf{A}_0 \tag{4.74}$$

and wave is transverse. $\omega = kc$, $k = |\mathbf{k}|$. Therefore $\mathbf{E}$ and $\mathbf{B}$ written as

$$\mathbf{E} = -2\omega A_0(\omega)\hat{e}\sin(\mathbf{k} \cdot \mathbf{r} - \omega t + \delta_\omega) \tag{4.75}$$

$$\mathbf{B} = -2A_0(\omega)(\mathbf{k} \times \hat{e}\sin(\mathbf{k} \cdot \mathbf{r} - \omega t + \delta_\omega) \tag{4.76}$$

where we have written $\mathbf{A}_0 = A_0(\omega)\hat{e}$. The direction of $\mathbf{E}$ is along $\hat{e}$, which specifies the polarization of the radiation, and is called the polarization vector.

Linearly polarized plane wave is a plane wave with its electric field vector $\mathbf{E}$ always in the direction of the polarization vector $\hat{e}$. $\mathbf{k} \perp \hat{e}$. In general, the polarization state of an electromagnetic wave may be given interms of two circularly polarized waves.

If we consider the vector potentials $\mathbf{A}^L(\omega; \mathbf{r}, t)$ and $\mathbf{A}^R(\omega; \mathbf{r}, t)$ defined by for a propogation along the z–direction

$$A_x^L = A_x^R = \sqrt{2}A_0(\omega)\cos(kz - \omega t + \delta_\omega) \tag{4.77}$$

$$A_y^L = -A_y^R = -\sqrt{2}A_0(\omega)\sin(kz - \omega t + \delta_\omega)$$

$$A_z^L = A_z^R = 0$$

The corresponding electric field vectors $\mathbf{E}^L$ and $\mathbf{E}^R$ are such that

$$E_x^L = E_x^R = -\sqrt{2}\omega A_0(\omega)\sin(kz - \omega t + \delta_\omega) \qquad (4.78)$$

$$E_y^L = -E_y^R = -\sqrt{2}\omega A_0(\omega)\cos(kz - \omega t + \delta_\omega)$$

$$E_z^L = E_z^R = 0$$

The radiation described by $\mathbf{E}^L$ is said to be left–hand circularly polarized and that corresponding to $\mathbf{E}^R$ is right–hand circularly polarized.

By forming the combination $a_L\mathbf{E}^L + a_R\mathbf{E}^L$ where a_L and a_R are complex coefficients, $|a_L|^2 + |a_R|^2 = 1$, radiation in any state of polarization can be produced. For example, if $a_L = a_R = 1$, we obtain linearly polarized radiation, with the electric field vector oriented along the x–axis.

The component of the spin of the radiation particle (photon) along the direction of the propagation is called the **helicity** of the particle. A photon with helicity $+\hbar$ is always left–hand circularly polarized and one with helicity $-\hbar$ is always right–hand circularly polarized.

4.6   Worked examples

Example - 4.1 :

Find the energy spectrum of a system whose Hamiltonian is

$$H = H_0 + H_1 = -\frac{\hbar^2}{2m}\frac{d^2}{dx^2} + \tfrac{1}{2}m\omega^2 x^2 + ax^3 + bx^4$$

where a and b are small constants (the anharmonic oscillator).

Solution :

If $a = b = 0$, the Hamiltonian H reduces to that of a linear harmonic oscillator, H_0 say. Let $H_1 = ax^3 + bx^4$ be the perturbation. The integral must be calculated

$$< \psi_n|H_1|\psi_n > = a < \psi_n|x^3|\psi_n > + b < \psi_n|x^4|\psi_n >$$

where ψ_n denotes the n th harmonic oscillator eigenfunction. Since x^3 is an odd function and ψ_n^2 is an even one, the first integral vanishes. The second integral is equal to

$$< \psi_n|x^4|\psi_n >= \int_{-\infty}^{+\infty} \psi_n x^4 \psi_n > dx = \tfrac{3}{4}x_0^4(2n^2 + 2n + 1) \quad , \quad x_0^2 = \tfrac{\hbar}{m\omega}$$

The first order perturbation of the energy level E_n^0 is thus given by

$$E_n^1 = \tfrac{3}{4}x_0^4(2n^2 + 2n + 1)$$

so that, to this approximation, the energy spectrum of the oscillator is

$$E_n = E_n^0 + E_n^1 = (n + \tfrac{1}{2})\hbar\omega + \tfrac{3}{4}x_0^4(2n^2 + 2n + 1).$$

Example - 4.2 :

A linear harmonic oscillator is perturbed by an electric field of strength $\mathcal{E}$. If the oscillating mass has charge $-e$, the perturbing Hamiltonian becomes $H_1 = e\mathcal{E}x$. Determine the perturbation correction to the energy through second order.

Solution :

The energy to second order is given by $E_n = E_n^{(0)} + E_n^{(1)} + E_n^{(2)}$.

$$E_n^{(1)} =< \psi_n^{(0)}|H_1|\psi_n^{(0)} >= e\mathcal{E} < n|x|n >= 0$$

$$E_n^{(2)} = \sum_{k \neq n} \frac{|<\psi_n^{(0)}|H_1|\psi_k^{(0)}>|^2}{E_n^{(0)} - E_k^{(0)}}$$

The matrix elements necessary for the second order correction are

$$< n|x|k >= \sqrt{\tfrac{n+1}{2\alpha}} \quad , \quad k = n+1 \quad ; \quad < n|x|k >= \sqrt{\tfrac{n}{2\alpha}} \quad , \quad k = n-1$$

here $\alpha = 2\pi\mu\nu/\hbar$, $\nu = (1/2\pi)\sqrt{k/\mu}$.

The energy to second order is then

$$E_n = (n + \tfrac{1}{2})h\nu + (e\mathcal{E})^2 \left[\frac{(\frac{n+1}{2a})}{-h\nu} + \frac{(\frac{n}{2a})}{h\nu} \right] = (n + \tfrac{1}{2})h\nu - \frac{e^2\mathcal{E}^2}{8\pi^2\mu\nu^2}$$

Thus the correction to second order is identical for all states.

Example - 4.3 :

A particle of mass m is constrained to move in the xy plane so that the Hamiltonian is

$$H = \tfrac{1}{2m}(p_x^2 + p_y^2) + \tfrac{1}{2}k(x^2 + y^2) + axy$$

Use the degenerate perturbation theory to determine the energy splitting for the lowest degenerate states and the first order wave functions for these states.

Solution :

The energies and degeneraties for the two–dimensional harmonic oscillator (when $a = 0$) are the following: $E = E_{n_x n_y} = (n_x + n_y + 1)h\nu$

E	n_x	n_y	g
$h\nu$	0	0	1
$2h\nu$	1	0	2
	0	1	
$3h\nu$	1	1	3
	2	0	
	0	2	
$\vdots$	$\vdots$	$\vdots$	$\vdots$

The degenerate functions corresponding to $2h\nu$ are

$$n_x = 1 \ , \quad n_y = 0 \quad \rightarrow \quad \phi_1 = \psi_1(x)\psi_0(y) = |1, 0>$$

$$n_x = 0 \ , \quad n_y = 1 \quad \rightarrow \quad \phi_2 = \psi_0(x)\psi_1(y) = |0, 1>$$

$\Psi_2 = C_1\phi_1 + C_2\phi_2$

When the perturbation $H_1 = axy$ is present the secular equation
$|(H_1)_{ij} - ES_{ij}| = 0$
must be solved to obtain the first order corrections to the energy. Here

$(H_1)_{ij} = < i|axy|j > = < n_x n_y|axy|n'_x n'_y > = a < n_x|x|n'_x > < n_y|y|n'_y >$

The secular equation becomes

$$\begin{vmatrix} a < 1|x|1 > < 0|y|0 > -E_2^{(1)} & a < 1|x|0 > < 0|y|1 > \\ \\ a < 0|x|1 > < 1|y|0 > & a < 0|x|0 > < 1|y|1 > -E_2^{(1)} \end{vmatrix} = 0$$

The nonvanishing matrix elements are:

$< 1|x|0 > = < 0|x|1 > = < 1|y|0 > = < 0|y|1 > = 1/\sqrt{2\alpha}$

where $\alpha = 4\pi^2 m\nu/h$, we obtain

$$\begin{vmatrix} -E_2^{(1)} & \frac{a}{2\alpha} \\ \\ \frac{a}{2\alpha} & -E_2^{(1)} \end{vmatrix} = 0 \quad , \quad E_{2\mp}^{(1)} = \mp\frac{a}{2\alpha}$$

The eigenfunctions are determined as follows:

$E_{2_+} = E_2^{(0)} + E_{2_+}^{(1)} \quad \rightarrow \quad -(\frac{a}{2\alpha})C_1 + (\frac{a}{2\alpha})C_2 = 0 \quad \rightarrow \quad C_1 = C_2$

$\psi_{2_+} = \frac{1}{\sqrt{2}}(|1,0 > +|0,1 >)$

$E_{2_-} = E_2^{(0)} + E_{2_-}^{(1)} \quad \rightarrow \quad (\frac{a}{2\alpha})C_1 + (\frac{a}{2\alpha})C_2 = 0 \quad \rightarrow \quad C_1 = -C_2$

$\psi_{2_-} = \frac{1}{\sqrt{2}}(|1,0 > -|0,1 >)$.

Example - 4.4 :

Suppose that $H_0 u_n = E_n u_n$ and $H = H_0 + H_1(t)$, where $H_1 \ll H_0$.

The solutions ψ of the time–dependent wave equation, $H\psi = i\hbar \frac{\partial \psi}{\partial t}$, can be expanded as follows:

$$\psi(x, t) = \sum_m a_m(t) e^{-iEt/\hbar} u_m(x)$$

Given the initial conditions, $a_n(0) = 1$, $a_k(0) = 0$ for $k \neq n$, show that

$$a_k(t) = -\tfrac{i}{\hbar} \int_0^t < k|H_1(t')|n > e^{i\omega_{kn}t'} dt',$$

where $\omega_{kn} = (E_k - E_n)/\hbar$. You may assume that $a_k(t) \ll 1$ for $k \neq n$. These initial conditions imply that we are dealing with first order perturbation theory.

Solution :

Substitution of ψ into the wave equation gives:

$$(H_0 + H_1(t)) \sum_m a_m e^{-iE_m t/\hbar} u_m$$

$$= i\hbar \left[\sum_m \dot{a}_m e^{-iE_m t/\hbar} u_m + \sum_m a_m (-\tfrac{i}{\hbar} E_m) e^{-iE_m t/\hbar} u_m \right]$$

Therefore, using $H_0 u_m = E_m u_m$ we finf :

$$\sum_m a_m H_1 u_m e^{-iE_m t/\hbar} = i\hbar \sum_m \dot{a}_m u_m e^{-iE_m t/\hbar}$$

Multiplication from left by u_k^* and integration then gives:

$$\sum_m a_m < k|H_1|m > e^{-iE_m t/\hbar} = i\hbar \dot{a}_k e^{-iE_k t/\hbar}$$

or $\dot{a}_k = -\tfrac{i}{\hbar} \sum_m a_m < k|H_1|m > e^{i\omega_{km}t}$ is found.

But $a_n(t) \sim 1$ and $a_m(t) \ll 1$, so that this equation can be approximated by setting $a_m = \delta_{nm}$ on the r.h.s. Finally, integration from 0 to t gives:

$$a_k(t) = -\tfrac{i}{\hbar} \int_0^t < k|H_1|n > e^{i\omega_{kn}t} dt \ .$$

Example - 4.5 :

Using the result of Example - 4.4 and assuming that $H_1(t) = 2V_0 \cos(\omega t)$ show that:

(a) The probability of finding a system in the state f at time t given that it was in the state i at $t = 0$ is

$$P_f(t) = |a_f|^2 = \frac{|V_{fi}|^2}{\hbar^2}\left(\frac{\sin^2(x)}{x^2}\right)t^2$$

where $V_{fi} = <f|V_0|i>$ and $x = (\omega_{fi} - \omega)t/2$.

(b) The transition rate from state i to state f (per system) can be written as

$$W_{fi} = \frac{P_f(t)}{t} = \frac{2\pi}{\hbar}|V_{fi}|^2 n(h\nu_{fi})$$

where $n(h\nu_{fi})$ is either the density of final states in the neighborhood of E_f, i.e., $n(h\nu) = dN(E)/dE$, or the density of modes in the radiation field in the neighborhood of $h\nu_{fi}$. The equation W_{fi} is known as Fermi's Golden Rule.

Solution :

(a) Using $a_f(t)$ from Example - 4.4 we have:

$$a_f(t) = -\frac{i}{\hbar}\int_0^t <f|V_0|i> e^{i\omega_{fi}t'}2\cos(\omega t')dt'$$

$$= -\frac{i}{\hbar}V_{fi}\int_0^t e^{i\omega_{fi}t'}(e^{i\omega t'} + e^{-i\omega t'})dt' = -\frac{i}{\hbar}V_{fi}\left[\frac{e^{i(\omega_{fi}+\omega)t}-1}{i(\omega_{fi}+\omega)} + \frac{e^{i(\omega_{fi}-\omega)t}-1}{i(\omega_{fi}-\omega)}\right]$$

for $\omega_{fi} \sim \omega$ the second term is the brackets above can be shown to dominate by use of l'Hospital's rule. This is associated with absorption of energy since $E_f - E_i \cong \hbar\omega$. Then keeping only the larger term and rearranging gives

$$a_f(t) = -\frac{i}{\hbar}V_{fi}e^{i(\omega_{fi}-\omega)t/2}\left[\frac{e^{i(\omega_{fi}-\omega)t/2}-e^{-i(\omega_{fi}-\omega)t/2}}{i(\omega_{fi}-\omega)t/2}\right]\frac{t}{2}$$

$$= -\tfrac{i}{\hbar} V_{fi} e^{i(\omega_{fi}-\omega)t/2} \left(\frac{\sin[(\omega_{fi}-\omega)t/2]}{[(\omega_{fi}-\omega)t/2]} \right) t$$

Finally, $P_f(t) = |a_f|^2 = \frac{|V_{fi}|^2}{\hbar^2} \left(\frac{\sin^2 x}{x^2} \right) t^2$, where $x = (\omega_{fi} - \omega)t/2$.

(b) The transition rate to a band of states close to E_f can be written as

$$W_{fi} = \tfrac{1}{t} \int_0^\infty n(E) P_f(t) dE = \frac{|V_{fi}|^2}{t\hbar^2} \int n(\hbar\omega_{fi}) \left(\frac{\sin^2 x}{x^2} \right) 2t\hbar dx$$

$$= \frac{2|V_{fi}|^2}{\hbar} n(h\nu_{fi}) \int \frac{\sin^2 x}{x^2} dx = \frac{2\pi}{\hbar} |V_{fi}|^2 n(h\nu_{fi})$$

For long times t, the function $\sin x/x$ is sharply peaked at $\omega = \omega_{fi}$ and the density of states function $n(h\nu_{fi})$ can safely be removed from the integral. The integral from tables is then

$$\int_{-\infty}^\infty \frac{\sin^2 x}{x^2} dx = \pi.$$

If $n(h\nu_{fi})$ were sharply peaked at E_f and t were short, it would be appropriate to replace $\sin x/x$ by unity. The rate would then not be independent of time.

Example - 4.6 :

Suppose that the probability of a transition from the ith excited state in the time interval Δt is proportional to Δt. Show that the population of the ith state depends on time as $N_i(t) = N_i(0)e^{-k_i t}$, where $N_i(0)$ is the population at $t = 0$ and k_i is the decay constant.

Solution :

Let the probability of a transition for one system in the time Δt be $k_i \Delta t$. Then the change in the population of the ith state in this time interval must be $\Delta N_i = -N_i k_i \Delta t$. In the limit that ΔN_i and Δt become infinitesimally small we can write $\frac{dN_i}{N_i} = -k_i dt$ which immediately gives the solution $N_i(t) = N_i(0)e^{-k_i t}$.

Chapter 5

Quantum Theory of One–Electron Atoms

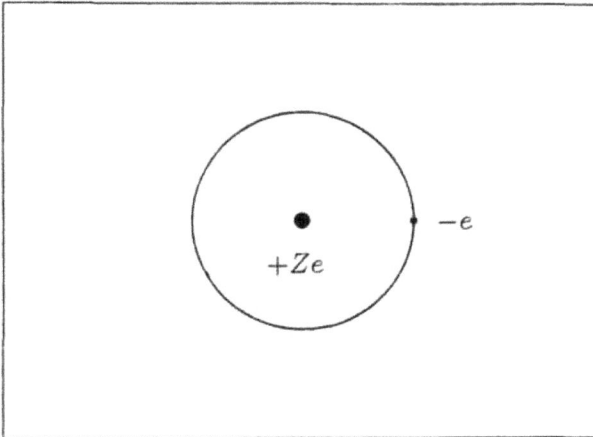

5.1   The Schrödinger equation for one–electron atoms

To apply the Schrödinger equation

$$-\frac{\hbar^2}{2m}\left(\frac{\partial^2}{\partial x^2} + \frac{\partial^2}{\partial y^2} + \frac{\partial^2}{\partial z^2}\right)\psi + V(x, y, z)\psi = E\psi \qquad (5.1)$$

to the hydrogen atom, we need to know the form of the potential energy $V(x, y, z)$.

The potential energy of an electron in the electrostatic field of a nucleus of charge Ze at a distance r is

$$V = -\frac{Ze^2}{r} = -\frac{Ze^2}{\sqrt{x^2 + y^2 + z^2}} \qquad (5.2)$$

The partial differential eqation, Eq.(5.1), with potential, Eq.(5.2), cannot be simply solved for the function $\psi(x, y, z)$ in cartesian coordinates because there is no way to separate the variables; that is, the solution cannot be expressed as a simple product function $\psi_x(x)\psi_y(y)\psi_z(z)$.

To simplify the solution, it is often helpful to transform it to a different coordinate system.

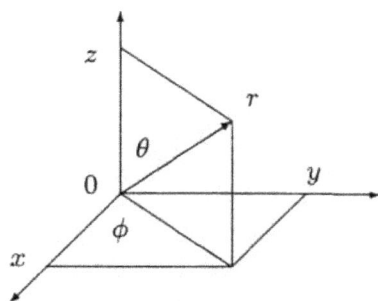

Fig. 5.1. Cartesian and spherical polar coordinates.

Since the potential energy, Eq.(5.2), is spherically symmetric, it is appropriate to introduce spherical polar coordinates:

$$x = r\sin\theta\cos\phi \ , \quad y = r\sin\theta\sin\phi \ , \quad z = r\cos\theta \qquad (5.3)$$

or

$$r = \sqrt{x^2 + y^2 + z^2} \quad , \quad \theta = \arccos\left(\frac{z}{\sqrt{x^2 + y^2 + z^2}}\right) \quad , \quad \phi = \arctan(y/x)$$

(5.4)

The resulting Schrödinger equation for the hydrogen atom in spherical polar coordinates is

$$-\frac{\hbar^2}{2m}\left[\frac{1}{r^2}\frac{\partial}{\partial r}\left(r^2\frac{\partial}{\partial r}\right) + \frac{1}{r^2 \sin\theta}\frac{\partial}{\partial\theta}\left(\sin\theta\frac{\partial}{\partial\theta}\right) + \frac{1}{r^2 \sin^2\theta}\frac{\partial^2}{\partial\phi^2}\right]\psi$$

$$-\frac{Ze^2}{r}\psi = E\psi \qquad (5.5)$$

where ψ is now considered to be a function of r, θ, and ϕ, and has the factored form

$$\psi(r,\theta,\phi) = R(r)Y(\theta,\phi) = R(r)\Theta(\theta)\Phi(\phi) \qquad (5.6)$$

5.2 The ground–state of one–electron atoms

To find the form of the functions we need to solve Eq.(5.4). The solution corresponding to the ground state of the hydrogen atom, we assume that ground state solution has no nodes. To eliminate the possibility of nodes in the angular functions Θ and Φ, we choose these functions to be constant, that is

$$\frac{\partial\psi}{\partial\theta} = 0 \quad and \quad \frac{\partial\psi}{\partial\phi} = 0 \qquad (5.7)$$

Writing $\psi(r,\theta,\phi) = R(r)$ we obtain the radial Schrödinger equation

$$-\frac{\hbar^2}{2m}\left(\frac{d^2R}{dr^2} + \frac{2}{r}\frac{dR}{dr}\right) - \frac{Ze^2}{r}R = ER \qquad (5.8)$$

A plot of $V(r)$ vs r shows that the region of lowest potential energy is near the origin, which suggests that the ground state wave function $R(r)$ should have a large amplitude at small r and should decrease rapidly as r increases.

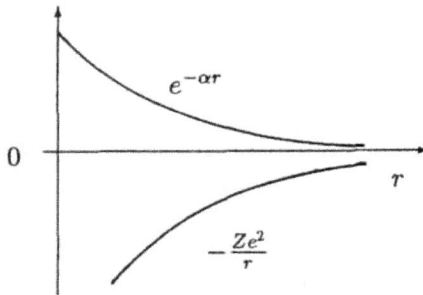

Fig. 5.2. Potential and wave function profiles.

The approximate equation of Eq.(5.8) for large r, called the asymptotic equation, can be written as

$$-\frac{\hbar^2}{2m}\frac{d^2R}{dr^2} = ER \quad (large\ r) \tag{5.9}$$

or

$$\frac{d^2R}{dr^2} = \alpha^2 R \tag{5.10}$$

where

$$\alpha^2 = -\frac{2mE}{\hbar^2} \tag{5.11}$$

The physically acceptable solution of Eq.(5.10) is

$$R(r) = Be^{-\alpha r} \tag{5.12}$$

$R(r) \rightarrow 0$ as $r \rightarrow \infty$. Substituting Eq.(5.12) into the Eq.(5.8) and rearranging, we obtain

$$\left(\alpha^2 + \frac{2mE}{\hbar^2}\right) + \frac{1}{r}\left(\frac{2mZe^2}{\hbar^2} - 2\alpha\right) = 0 \tag{5.13}$$

This equation, Eq.(5.13), can hold for all values of r if each of the terms in parantheses to be separately equal to zero. Eq.(5.13) can be solved for α; it yields

$$\alpha = \frac{Zme^2}{\hbar^2} = \frac{Z}{a_0} \tag{5.14}$$

When this value of α is inserted into Eq.(5.11), E is found to be

$$E = -\frac{Z^2\hbar^2}{2ma_0} = -\frac{\hbar^2}{me^2}\frac{Z^2e^2}{2a_0^2} = -\frac{Z^2e^2}{2a_0} \tag{5.15}$$

The energy given by Eq.(5.15) is identical with the Bohr theory result for the ground state. The ground state wave function ψ_1 is found to be

$$\psi_1 = Ne^{-Zr/a_0} \tag{5.16}$$

where N is a normalization constant. The criterion for determining the value of N is that the probability of finding the electron somewhere in space be unity,

$$\int \psi_1^2 dv = N^2 \int e^{-2Zr/a_0} dv = 1 \tag{5.17}$$

The differential volume element dv has the form (in spherical polar coordinates)

$$dv = r^2 \sin\theta dr d\theta d\phi \tag{5.18}$$

The limits of the variables are

$$0 \le r \le \infty \quad , \quad 0 \le \theta \le \pi \quad , \quad 0 \le \phi \le 2\pi \tag{5.19}$$

The normalization integral, Eq.(5.17), for the ground state wave function is

$$N^2 \int_0^{2\pi} d\phi \int_0^{\pi} \sin\theta d\theta \int_0^{\infty} e^{-2Zr/a_0} r^2 dr = 1 \tag{5.20}$$

Solving for N, we find

$$N = \left(\frac{Z^3}{\pi a_0^3}\right)^{1/2} \tag{5.21}$$

The normalized wave function is therefore

$$\psi_1 = \left(\frac{Z^3}{\pi a_0^3}\right)^{1/2} e^{-Zr/a_0} \tag{5.22}$$

For very accurate results, the electron mass m should be replaced by the reduced mass μ of the nucleus–electron system. For heavy atoms or molecules this nuclear–mass correction is usually omitted.

5.3   Ground–state wave function and probability

Let us now examine the spatial distribution of an electron in the ground state of the hydrogen atom $(Z = 1)$.

$\psi_1^2(r)$ gives the probability per unit volume of finding the electron at a point in space at a distance r from the nucleus.

The quantity

$$P_1(r) = 4\pi r^2 \psi_1^2(r) \tag{5.23}$$

is a probability distribution (radial distribution function) for the variable r; it is the probability per unit distance of observing the electron at a distance r from the origin.

The most probable value for r corresponds to the point for which the distribution $P_1(r)$ has its maximum.

$$\frac{dP_1(r)}{dr} = 0 \qquad gives \qquad r = a_0 \tag{5.24}$$

Thus, the most probable value of r is equal to the Bohr radius. This is the quantum theory analog to the Bohr picture of the ground state. Since the probability distribution is nonzero for all $r > 0$, the **size** of atoms is not well defined by the quantum theory model. One possible definition of size is in terms of the maximum of $P(r)$. An alternate possibility is to use the average or expectation value of r

$$< r >= \int \psi_1(r) r \psi_1(r) dv = 4\pi \left(\frac{Z^3}{\pi a_0^3} \right) \int_0^\infty e^{-2Zr/a_0} r^3 dr = \frac{3}{2} \frac{a_0}{Z} \tag{5.25}$$

for hydrogen $(Z = 1)$, $< r >= 3a_0/2$. Thus, the average value of r is somewhat larger than the most probable value.

Another useful limit is that obtained by finding the value $r_{0.9}$ such that there is a 90% probability that the $r \le r_{0.9}$. To determine $r_{0.9}$, we consider the radial integral

$$4\pi \left(\frac{Z^3}{\pi a_0^3} \right) \int_0^{r_{0.9}} e^{-2Zr/a_0} r^2 dr = 0.9 \tag{5.26}$$

solving for $r_{0.9}$, one finds $r_{0.9} \cong 2.6a_0$.

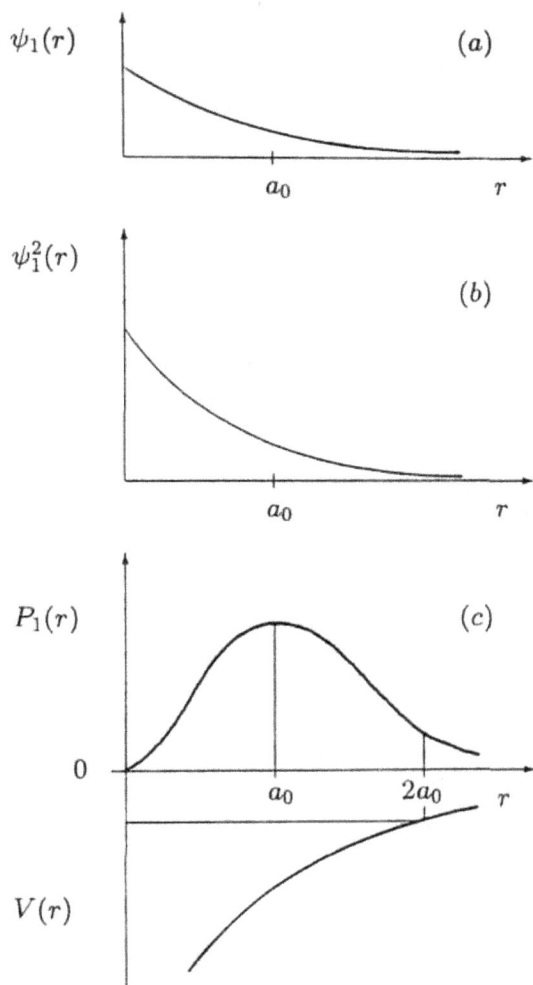

Fig. 5.3. (a) Wave function; (b) probability density; (c) probability distribution with potential energy.

5.4   Tunnel effect

Classically, a particle has a turning point where $V = E$. Since $K = 0$ at such a point, the classical particle is expected to be turned around or reflected by the potential barrier. For an electron in the hydrogen atom ground state, such a classical turning point occurs where

$$V(r) = -\frac{e^2}{r} = E = -\frac{e^2}{2a_0} \tag{5.27}$$

that is, at $r = 2a_0$. Although the probability distribution $P(r)$ has a nonzero value for $r > 2a_0$, the electron has access to a region which is forbidden by classical theory. Such penetration or tunneling into or through potential barrier is typical of quantum theory results.

5.5   Spherical excited states of $1 - e^-$ atoms

For the ground state wave function of hydrogen we assume that the angular functions Θ and Φ were constant, so that there would be no angular nodes; also the radial function $R(r)$ was nodeless.

For the first excited state, it is reasonable to suppose that the solution to Eq.(5.8) has a single node. We take the angular functions constant and solve Eq.(5.8) for $R(r)$. An appropriate function is a first–degree polynomial in r multiplied by an exponential function; that is,

$$\psi_2(r) = N_2(b - r)e^{-ar} \tag{5.28}$$

The factor $(b - r)$ gives a node at $r = b$. The constants a and b, and the energy E are evaluated in the same way as for the ground state. The function given in Eq.(5.28) is substituted for $R(r)$ in Eq.(5.8), the coefficients of the different powers of r are collected, and they are separately set equal to zero, because Eq.(5.8) must hold for all r. The results obtained are

$$a = \frac{Z}{2a_0} \quad , \quad b = \frac{2a_0}{Z} \quad , \quad and \quad E_2 = -\frac{1}{4}\frac{Z^2e^2}{2a_0} \tag{5.29}$$

and the normalized wave function can be written

$$\psi_2(r) = \frac{1}{4}\left(\frac{Z^3}{2\pi a_0^3}\right)^{1/2}(2 - Zr/2a_0)e^{-Zr/2a_0} \tag{5.30}$$

The same procedure can be repeated for states $2, 3, 4, ...$ nodes in the radial function $R(r)$ and constant angular functions. In general, if $(n - 1)$ is the number of nodes, the wave functions for spherically symmetric states (no θ or ϕ dependence) can be written

$$\psi_n(r) = N_n \left(\sum_{i=0}^{n-1} b_i r^i \right) e^{-ar} \tag{5.31}$$

where n is an integer which is called the **principal quantum number**; it can take on the values $n = 1, 2, 3, ...$ and is associated with a function that has $(n - 1)$ nodes, all of which are radial for the spherically symmetric solutions. For the nodeless ground state, $n = 1$.

For $n = 1$ and $n = 2$, we have seen that $E_1 = -Z^2 e^2 / 2a_0$ and $E_2 = -(1/4) Z^2 e^2 / 2a_0$, respectively. In the general case, it can be shown that

$$E_n = -\frac{1}{n^2} \frac{Z^2 e^2}{2a_0} \quad , \quad n = 1, 2, 3, ... \tag{5.32}$$

For spherically symmetric states of the hydrogen atom, the energy levels predicted by the Bohr model and the quantum theory are identical.

5.6 Functions without spherical symmetry

The possibility now arises for nodes in the angular functions $\Theta(\theta)$ and $\Phi(\phi)$, since they are no longer constant for states that are not spherically symmetric. As the simplest function of this type, consider those with one angular node. They have the form

$$xf(r) \quad , \quad yf(r) \quad , \quad zf(r)$$

or

$$r \sin\theta \cos\phi f(r) \quad , \quad r \sin\theta \sin\phi f(r) \quad , \quad r \cos\theta f(r) \tag{5.33}$$

with $f(r)$ to be determined by substitution into the Schrödinger equation. These three functions are equivalent except for their orientation, and each has one angular node. Consider $xf(r)$ as an example. If we substitute $xf(r)$ into Eq.(5.5), we obtain

$$-\frac{\hbar^2}{2m} \left(\frac{d^2 f(r)}{dr^2} + \frac{4}{r} \frac{df(r)}{dr} \right) - \frac{Ze^2}{r} f(r) = Ef(r) \tag{5.34}$$

The identical equation is obtained from the function $yf(r)$ and $zf(r)$.

The form of Eq.(5.34) is very similar to the previous radial equation, Eq.(5.8). We try the same exponential–times–polynomial form for $f(r)$. Since we are interested in the lowest state with one angular node, we assume a simple nodeless exponential, that is $f(r) = e^{-ar}$ with a to be determined from Eq.(5.34). The result of setting the coefficient of each power of r equal to zero is

$$a = \frac{Z}{2a_0} \quad , \quad E = -\frac{1}{4}\frac{Z^2 e^2}{2a_0} \tag{5.35}$$

Evaluation of the normalization constant yields

$$\psi_x = \frac{1}{4}\left(\frac{Z^5}{2\pi a_0^5}\right)^{1/2} r\sin\theta\cos\phi\, e^{-Zr/2a_0} \tag{5.36}$$

$$\psi_y = \frac{1}{4}\left(\frac{Z^5}{2\pi a_0^5}\right)^{1/2} r\sin\theta\sin\phi\, e^{-Zr/2a_0}$$

$$\psi_z = \frac{1}{4}\left(\frac{Z^5}{2\pi a_0^5}\right)^{1/2} r\cos\theta\, e^{-Zr/2a_0}$$

The three simplest angular functions all have the same energy (degenerate), which is also equal to the energy of ψ_2, the second spherically symmetric solution.

In the hydrogen atom, the symmetry is provided by the spherically symmetric potential. The spherically symmetric function ψ_2 has the same energy as the angle–dependent functions ψ_x, ψ_y, ψ_z is not due to simple spatial symmetry; instead it is a consequence of the Coulomb interaction between the electron and the nucleus.

Additional non–spherically symmetric functions can be found by using the same angular dependence and introducing one more radial nodes. Alternatively, the number of angular nodes can be increased; for example, functions such as $xyg(r)$, $xzg(r)$ can be introduced and $g(r)$ determined so as to satisfy the Schrödinger equation.

5.7 Quantum numbers for hydrogen–like atoms

For hydrogen–like atoms, three quantum numbers are required to completely specify the state or wave function. The first one is the principle quantum number n, which determines the energy, Eq.(5.32), and gives the total number of nodes, both radial and angular, as $(n-1)$.

A second quantum number for the hydrogen atom is designated by the letter ℓ and it specifies the number of angular nodes. It is often called the **azimuthal quantum number**. The allowed values of ℓ are

$$\ell = 0, 1, 2, ..., (n-1) \tag{5.37}$$

The number of angular nodes must be less than or equal to the total number of nodes. There are thus ℓ angular nodes and $n - \ell - 1$ radial nodes, $(n_r = n - \ell - 1$; radial quantum number).

The spherically symmetric functions have $\ell = 0$ (no angular nodes). The functions ψ_x, ψ_y, ψ_z all have $\ell = 1$ (one angular node).

The azimuthal quantum number specifies the magnitude squared, L^2, of the angular momentum of the electron in the atom.

$$L^2 = \ell(\ell+1)\hbar^2 \quad , \quad \ell = 0, 1, 2, ..., (n-1) \tag{5.38}$$

The electronic angular momentum of the hydrogen atom is quantized.

In the Bohr theory the motion of the electron was planar, in the quantum theory planar motion is not allowed. For nonplanar motion, the angular momentum is a vector with components L_x, L_y, L_z.

By an appropriate choice of axes, L_z can be made to coincide with the angular momentum ($p_\phi = n\hbar$) for the planar motion of the Bohr theory. We expect the same quantization rules to apply to L_z and p_ϕ. This is true in that L_z is given by the expression

$$L_z = m_\ell \hbar \tag{5.39}$$

where the quantum number m_ℓ is a positive or negative integer or zero. The sign of m_ℓ determines whether the z–component of angular momentum points in the positive or negative z–direction. m_ℓ may be measured by observing the spectrum of the atom in a magnetic field; for this reason, m_ℓ is called the **magnetic quantum number**.

The inequality $|m_\ell| \leq \sqrt{\ell(\ell+1)}$ holds, and both m_ℓ and ℓ are integers. The allowed values of m_ℓ are

$$m_\ell = 0, \mp 1, \mp 2, ..., \mp \ell \tag{5.40}$$

States with different ℓ and m_ℓ but with the same n are degenerate. For a given n, there are n different possible values of ℓ. For each value of ℓ, there are $2\ell + 1$ different values of m_ℓ. The total number of states with the same energy (that is, the degeneracy) is

$$g = \sum_{\ell=0}^{n-1} (2\ell + 1) = n^2 \tag{5.41}$$

This n^2–fold degeneracy for the hydrogen–like atoms is due to the fact that $V(r) = -Ze^2/r$. A change in the form of the potential energy to a non–Coulombic interaction makes the energy depend on both n and ℓ. If in addition the spherical symmetry of the potential is removed by applying an external magnetic field, the energy depends on m_ℓ as well.

5.8   Wave functions for one–electron atoms

In terms of the quantum numbers n, ℓ, and m_ℓ, the wave functions for a one–electron atom with nuclear charge Ze can be specified as

$$\psi_{n\ell m_\ell}(r, \theta, \phi) = R_{n\ell}(r)\Theta_{\ell m_\ell}(\theta)\Phi_{m_\ell}(\phi) = R_{n\ell}(r)Y_{\ell m_\ell}(\theta, \phi) \tag{5.42}$$

where $Y_{\ell m_\ell}(\theta, \phi)$ is the spherical harmonics. These wave functions are normalized, that is,

$$\int \psi_{n\ell m_\ell}^2(r, \theta, \phi) r^2 dr \sin\theta d\theta d\phi = 1 \tag{5.43}$$

and they are orthogonal, that is,

$$\int \psi_{n\ell m_\ell} \psi_{n'\ell'm_\ell'} r^2 dr \sin\theta d\theta d\phi = \delta_{nn'}\delta_{\ell\ell'}\delta_{m_\ell m_\ell'} \tag{5.44}$$

If the wave functions are expressed in complex form, usually ϕ dependent part $\Phi_{m_\ell}(\phi)$, then $|\psi|^2 = \psi\psi^*$ is used in the integrals. These one–electron atomic wave functions are commonly called as **atomic orbitals**, and they are named according to the value of ℓ as follows:

$$\ell = 0(s) \,, \; 1(p) \,, \; 2(d) \,, \; 3(f) \,, \; 4(g) \,, \; 5(h) \, ...$$

Table 5.1: States of one–electron atoms (only for $n = 1$ and $n = 2$):

n ℓ $|m_\ell|$	spect. desig.	E_n $\times(e^2/2a_0)$	g	$\psi_{n\ell m_\ell}(r,\theta,\phi)$
1 0 0	$1s$	-1	1	$N_1 e^{-r'}$
2 0 0	$2s$	$-1/4$	4	$N_2(2 - r')e^{-r'/2}$
2 1 0	$2p_z$	$-1/4$		$N_2 r' e^{-r'/2}\cos\theta$
2 1 1, cos	$2p_x$	$-1/4$		$N_2 r' e^{-r'/2}\sin\theta\cos\phi$
2 1 1, sin	$2p_y$	$-1/4$		$N_2 r' e^{-r'/2}\sin\theta\sin\phi$

$$r' = Zr/a_0 \,, \; N_1 = (Z^3/(\pi a_0^3))^{1/2} \,, \; N_2 = N_1/\sqrt{32}$$

The general form of the Hamiltonian for a one–particle system with time–independent, central potential $(V(r))$ is given by

$$H = -\frac{\hbar^2}{2m}\nabla^2 + V(r) \tag{5.45}$$

$$= -\frac{\hbar^2}{2m}\left[\frac{1}{r^2}\frac{\partial}{\partial r}\left(r^2\frac{\partial}{\partial r}\right) + \frac{1}{r^2\sin\theta}\frac{\partial}{\partial\theta}\left(\sin\theta\frac{\partial}{\partial\theta}\right) + \frac{1}{r^2\sin^2\theta}\frac{\partial^2}{\partial\phi^2}\right] + V(r)$$

we may also write this Hamiltonian as

$$H = -\frac{\hbar^2}{2m}\left[\frac{1}{r^2}\frac{\partial}{\partial r}\left(r^2\frac{\partial}{\partial r}\right) - \frac{L^2}{\hbar^2 r^2}\right] + V(r) \tag{5.46}$$

where L is the angular momentum operator,

$$\mathbf{L} = L_x\mathbf{i} + L_y\mathbf{j} + L_z\mathbf{k} \tag{5.47}$$

which operates on angular variables only.

Thus, the general form of the time–independent Schrödinger equation is given by

$$\left\{ -\frac{\hbar^2}{2m} \left[\frac{1}{r^2} \frac{\partial}{\partial r} \left(r^2 \frac{\partial}{\partial r} \right) - \frac{L^2}{\hbar^2 r^2} \right] + V(r) \right\} \psi_{n\ell m_\ell}(\mathbf{r}) = E\psi_{n\ell m_\ell}(\mathbf{r}) \quad (5.48)$$

The operators H, L^2, L_z commute each other, that is,

$$[H, L^2] = [H, L_z] = [L^2, L_z] = 0 \quad (5.49)$$

Commutation relation is defined as

$$[A, B] = AB - BA \quad (5.50)$$

Physically this means that the same function (wave function) is the eigenfunction of these operators, namely,

$$H\psi_{n\ell m_\ell}(\mathbf{r}) = E_n \psi_{n\ell m_\ell}(\mathbf{r}) \quad (5.51)$$

$$L^2 \psi_{n\ell m_\ell}(\mathbf{r}) = \ell(\ell+1)\hbar^2 \psi_{n\ell m_\ell}(\mathbf{r}) \quad (5.52)$$

$$L_z \psi_{n\ell m_\ell}(\mathbf{r}) = m_\ell \hbar \psi_{n\ell m_\ell}(\mathbf{r}) \quad (5.53)$$

If we write

$$\psi_{n\ell m_\ell}(\mathbf{r}) = \psi_{n\ell m_\ell}(r, \theta, \phi) = R_{n\ell}(r) Y_{\ell m_\ell}(\theta, \phi) \quad (5.54)$$

then the general form of the radial Schrödinger equation is given by

$$\left\{ -\frac{\hbar^2}{2m} \left[\frac{1}{r^2} \frac{d}{dr} \left(r^2 \frac{d}{dr} \right) - \frac{\ell(\ell+1)}{r^2} \right] + V(r) \right\} R_{n\ell}(r) = E_n R_{n\ell}(r) \quad (5.55)$$

The term with ℓ on the left hand side represents the centrifugal barrier potential, and $V(r)$ is the Coulomb potential.

5.9 The Virial Theorem for one–electron atoms

In classical planetary model a bound particle with electrostatic interaction potential obeys the Virial Theorem, namely,

$$E = \frac{1}{2}\bar{V} = -\bar{T} \quad (5.56)$$

where $\bar{V}$ and $\bar{T}$ are time averages of the potential and kinetic energies, respectively. In the quantum theory, the Virial Theorem for the hydrogen atom takes the analogous form

$$E_n = \frac{1}{2} < V >_{n\ell m_\ell} = - < T >_{n\ell m_\ell} \qquad (5.57)$$

where $< V >$ and $< T >$ are the quantum mechanical average values,

$$< V >_{n\ell m_\ell} = - \int_0^\infty \int_0^{2\pi} \int_0^\pi \psi_{n\ell m_\ell} \left(\frac{Ze^2}{r} \right) \psi_{n\ell m_\ell} r^2 \sin\theta \, dr \, d\theta \, d\phi \qquad (5.58)$$

$$< T >_{n\ell m_\ell} = - \int_0^\infty \int_0^{2\pi} \int_0^\pi \psi_{n\ell m_\ell} \left(\frac{\hbar^2}{2m} \nabla^2 \right) \psi_{n\ell m_\ell} r^2 \sin\theta \, dr \, d\theta \, d\phi \qquad (5.59)$$

For the ground state

$$< V >_{100} = - \frac{Z^2 e^2}{a_0} \quad , \quad < T >_{100} = + \frac{Z^2 e^2}{2a_0} \qquad (5.60)$$

For a general state $n\ell m_\ell$, we obtain

$$< V >_{n\ell m_\ell} = 2E_n = - \frac{Z^2 e^2}{n^2 a_0} \quad , \quad < T >_{n\ell m_\ell} = -E_n = \frac{Z^2 e^2}{2n^2 a_0} \qquad (5.61)$$

Expectation values of some powers of r in hydrogenic systems:

$$< r >_{n\ell m_\ell} = \frac{a_0 n^2}{Z} \left[1 + \frac{1}{2} \left\{ 1 - \frac{\ell(\ell+1)}{n^2} \right\} \right] \qquad (5.62)$$

$$< \frac{1}{r} >_{n\ell m_\ell} = \frac{Z}{a_0 n^2} \qquad (5.63)$$

$$< r^2 >_{n\ell m_\ell} = \frac{a_0^2 n^4}{Z^2} \left[1 + \frac{3}{2} \left\{ 1 - \frac{\ell(\ell+1) - \frac{1}{3}}{n^2} \right\} \right] \qquad (5.64)$$

$$< \frac{1}{r^2} >_{n\ell m_\ell} = \frac{Z^2}{a_0^2 n^3 (\ell + \frac{1}{2})} \qquad (5.65)$$

$$< r^3 >_{n\ell m_\ell} = \frac{a_0^3 n^6}{Z^3} \times$$

$$\left\{ 1 + \frac{27}{8} \left[1 - \frac{1}{n^2} \left(\frac{35}{27} - \frac{10}{9}(\ell+2)(\ell-1) \right) + \frac{1}{9n^4}(\ell+2)(\ell+1)\ell(\ell-1) \right] \right\} \tag{5.66}$$

$$< \frac{1}{r^3} >_{n\ell m_\ell} = \frac{Z^3}{a_0^3 n^3 \ell(\ell + \frac{1}{2})(\ell+1)} \tag{5.67}$$

The recursion formula for evaluating $< r^m >_{n\ell m_\ell}$ was given by Kramers:

$$Z^2 \left(\frac{m+1}{n^2} \right) < r^m >_{n\ell m_\ell} - (2m+1)Z < r^{m-1} >_{n\ell m_\ell}$$

$$+ m(\ell + \frac{1}{2} + \frac{1}{2}m)(\ell + \frac{1}{2} - \frac{1}{2}m) < r^{m-2} >_{n\ell m_\ell} = 0 \tag{5.68}$$

5.10 Geometrical details of hydrogen–like orbitals

In the quantum–mechanical description of the atom, the orbits are replaced with the probability distribution functions for the coordinates of the electron. An electron in an atom is thus represented by a probability cloud whose density, ψ^2, varies from point to point. Different states of the electron (corresponding to different wave function) have distributions with different shapes and density patterns. Plots of the radial factor $R(r)$ and angular factor $\Theta\Phi = Y$ can be used to give pictorial representations of the electronic wave functions for the various states.

The radial functions $R_{n\ell}(r)$ have been normalized so that

$$\int_0^\infty (R_{n\ell}(r))^2 r^2 dr = 1 \tag{5.69}$$

The number of radial nodes is $n_r = n - \ell - 1$. Only the s functions ($\ell = 0$) have nonzero values at the origin. The s orbital has no angular dependence. p_z is independent of ϕ, the angular parts of the p functions have a single nodal plane, since $\ell = 1$.

Table 5.2: Possible atomic orbitals for $n = 1, 2, 3$:

ℓ n	0	1	2
1	$(1s)$		
2	$(2s)$	$(2p_x, 2p_y, 2p_z)$	
3	$(3s)$	$(3p_x, 3p_y, 3p_z)$	$(3d_{xy}, 3d_{xz}, 3d_{yz}, 3d_{x^2-y^2}, 3d_{3z^2-r^2})$

The relations between real and complex atomic orbitals:

$$p_z = p_0 = \sqrt{\frac{3}{4\pi}} \cos\theta \qquad\qquad p_0 = p_z$$

$$p_x = \frac{1}{\sqrt{2}}(p_{+1} + p_{-1}) = \sqrt{\frac{3}{4\pi}} \cos\theta\sin\phi \qquad p_1 = \frac{1}{\sqrt{2}}(p_x + ip_y)$$

$$p_y = \frac{1}{\sqrt{2}i}(p_{+1} - p_{-1}) = \sqrt{\frac{3}{4\pi}} \sin\theta\sin\phi \qquad p_{-1} = \frac{1}{\sqrt{2}}(p_x - ip_y)$$

$$d_{x^2-y^2} = \frac{1}{\sqrt{2}}(d_{+2} + d_{-2}) \qquad\qquad d_{+2} = \frac{1}{\sqrt{2}}(d_{x^2-y^2} + id_{xy})$$

$$d_{xy} = \frac{1}{\sqrt{2}i}(d_{+2} - d_{-2}) \qquad\qquad d_{-2} = \frac{1}{\sqrt{2}}(d_{x^2-y^2} - id_{xy})$$

$$d_{xz} = \frac{1}{\sqrt{2}}(d_{+1} + d_{-1}) \qquad\qquad d_{+1} = \frac{1}{\sqrt{2}}(d_{xz} + id_{yz})$$

$$d_{yz} = \frac{1}{\sqrt{2}i}(d_{+1} - d_{-1}) \qquad\qquad d_{-1} = \frac{1}{\sqrt{2}}(d_{xz} - id_{yz})$$

$$d_{z^2} = d_0 \qquad\qquad d_0 = d_{z^2}$$

Table 5.3: Atomic orbitals for $n = 1, 2, 3$:

n	ℓ	m_ℓ	orbital
1	0	0	$1s$
2	0	0	$2s$
2	1	0	$2p_z$
2	1	± 1	$2p_x(\cos), 2p_y(\sin)$
3	0	0	$3s$
3	1	0	$3p_z$
3	1	± 1	$3p_x(\cos), 3p_y(\sin)$
3	2	0	$3d_{3z^2-r^2}$
3	2	± 1	$3d_{xz}(\cos), 3d_{yz}(\sin)$
3	2	± 2	$3d_{x^2-y^2}(\cos), 3d_{xy}(\sin)$

Table 5.4: Normalized spherical harmonics, $Y_{\ell m_\ell}(\theta, \phi)$:

orbital	ℓ	m_ℓ	$Y_{\ell m_\ell}(\theta, \phi)$
s	0	0	$\frac{1}{\sqrt{4\pi}}$
p_0	1	0	$\sqrt{\frac{3}{4\pi}} \cos\theta$
$p_{\pm 1}$	1	± 1	$\sqrt{\frac{3}{8\pi}} \sin\theta e^{\pm i\phi}$
d_0	2	0	$\sqrt{\frac{5}{16\pi}}(3\cos^2\theta - 1)$
$d_{\pm 1}$	2	± 1	$\sqrt{\frac{15}{8\pi}} \cos\theta \sin\theta e^{\pm i\phi}$
$d_{\pm 2}$	2	± 2	$\sqrt{\frac{15}{32\pi}} \sin^2\theta e^{\pm 2i\phi}$

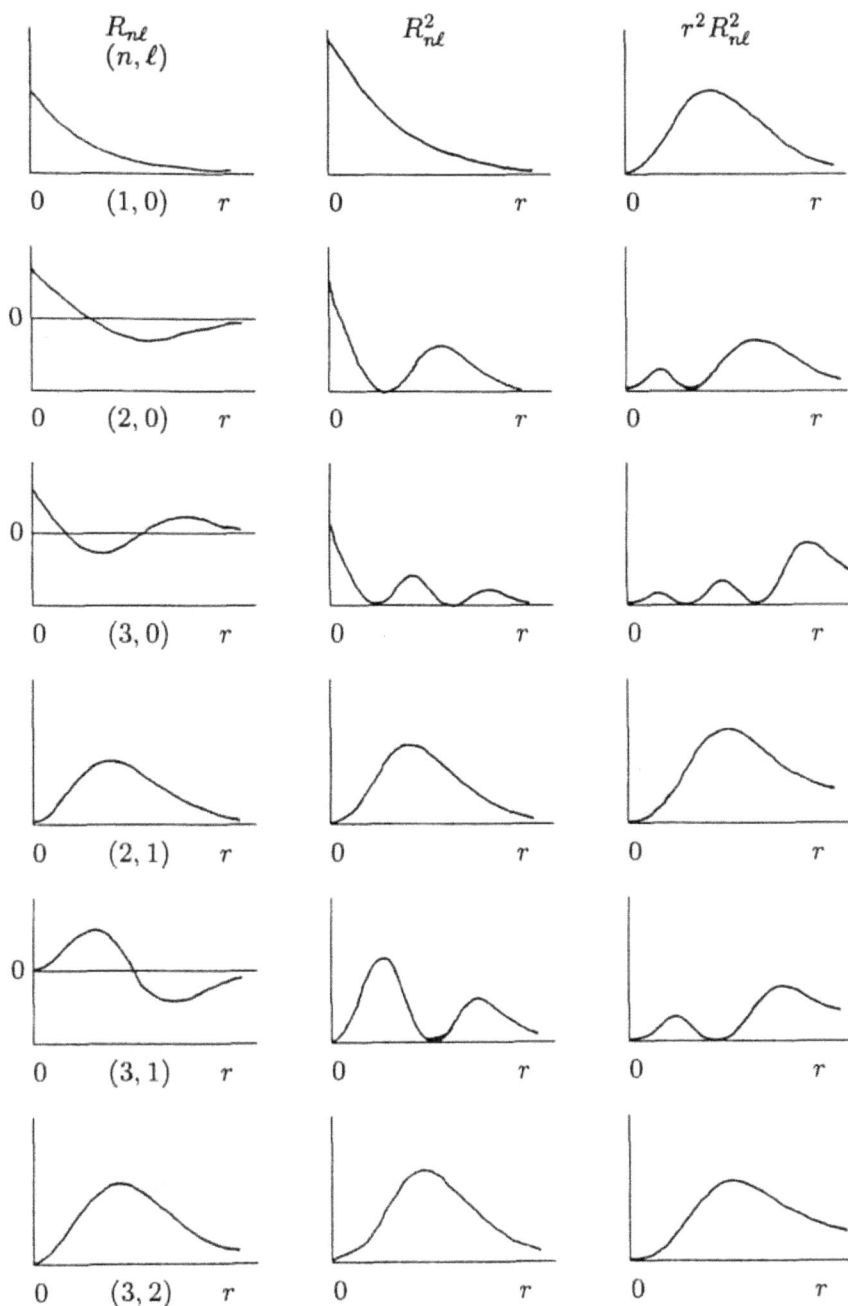

Fig. 5.4. Radial wave functions for hydrogenic atoms.

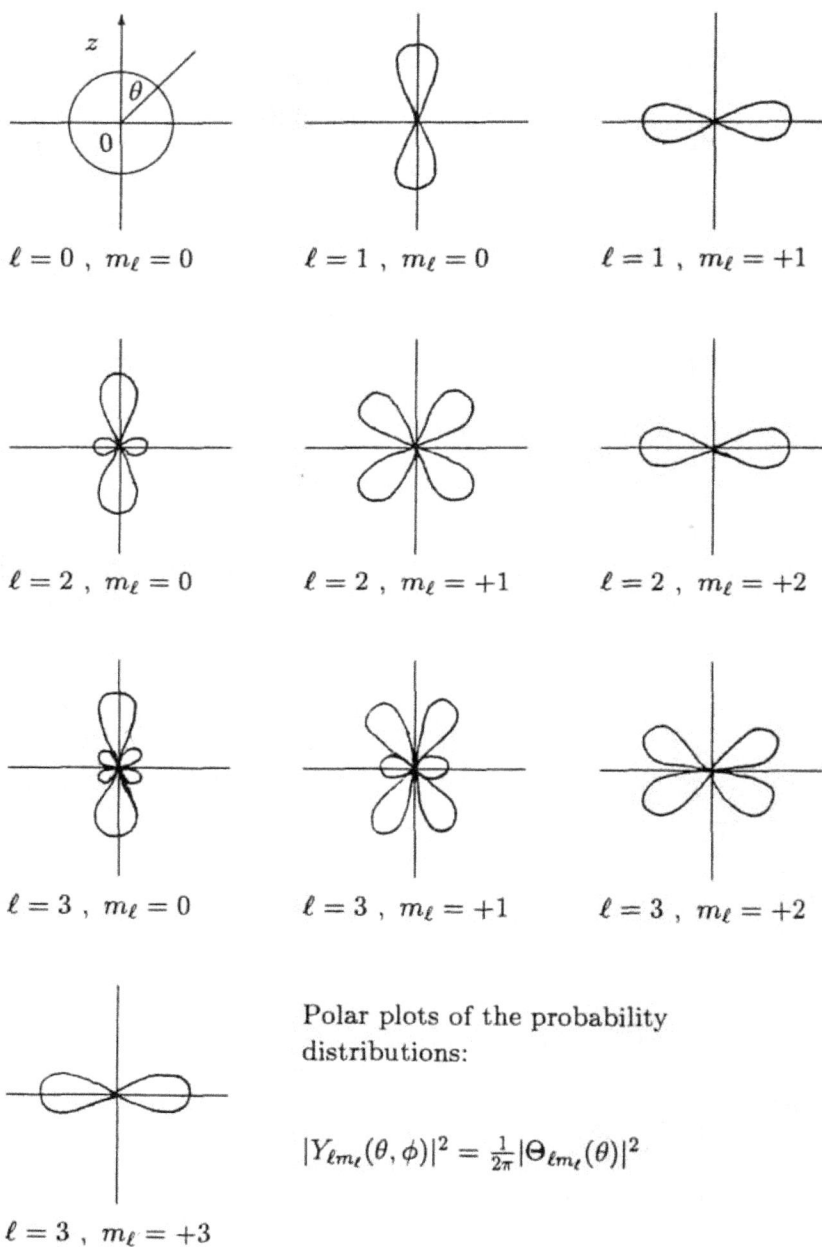

$\ell = 0$, $m_\ell = 0$ $\ell = 1$, $m_\ell = 0$ $\ell = 1$, $m_\ell = +1$

$\ell = 2$, $m_\ell = 0$ $\ell = 2$, $m_\ell = +1$ $\ell = 2$, $m_\ell = +2$

$\ell = 3$, $m_\ell = 0$ $\ell = 3$, $m_\ell = +1$ $\ell = 3$, $m_\ell = +2$

$\ell = 3$, $m_\ell = +3$

Polar plots of the probability distributions:

$$|Y_{\ell m_\ell}(\theta, \phi)|^2 = \frac{1}{2\pi}|\Theta_{\ell m_\ell}(\theta)|^2$$

Fig. 5.5. Angular wave functions for hydrogenic atoms.

5.11 Energy levels and spectrum of the hydrogen atom

Once the various possible states of the hydrogen atom and the corresponding energy levels are found by solving the Schrödinger equation, the Bohr frequency rule can be used to determine the wave number of the spectral line corresponding to transitions from one level to another. Transitions can take place between states with any two values of n. However, it is found that only those transitions are allowed for which ℓ changes by ± 1. The **selection rules** for the most common spectral transitions due to the interactions between the electron and the dipole electric field of the radiation are

$$n_1 - n_2 = \Delta n \quad arbitrary$$

$$\ell_1 - \ell_2 = \Delta \ell = \pm 1 \tag{5.70}$$

$$m_{\ell_1} - m_{\ell_2} = \Delta m_\ell = 0 \, , \, \pm 1$$

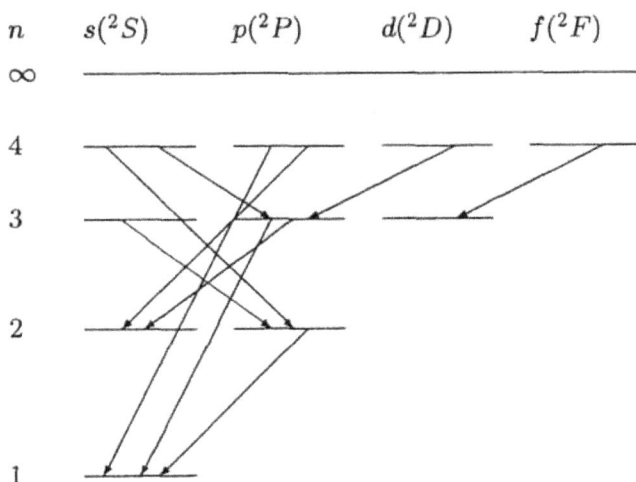

Fig. 5.6. Allowed transitions for hydrogenic atoms.

To determine the selection rules and the intensities of the spectral lines, we calculate the probability of finding the system in one state (say, the excited state, ψ_{n_2}) at time t, given that the system was in another state (say, the ground state, ψ_{n_1}) at time zero.

For a hydrogen–like atom interacting with radiation, the potential energy function includes the terms representing the interaction of the electron with the electromagnetic wave. The most important of these is the **electric dipole** term. For example, for a wave with the electric field oriented in the x–direction, the electric dipole term is $-exE_x$.

Writing the electric field of a plane wave moving in the z–direction, we have

$$V(x, y, z, t) = -\frac{Ze^2}{r} - exE_x^0 \cos[2\pi(z/\lambda - \nu t)] \tag{5.71}$$

After solving the time–dependent Schrödinger equation, we obtain the probability of finding the atom in the state ψ_{n_2} at time t as

$$P_{n_2}(t) \cong \frac{e^2(E_x^0)^2}{\hbar^2} | < \psi_{n_2}|x|\psi_{n_1} > |^2 t^2 \tag{5.72}$$

The intensity $I_{n_1 \to n_2}$ of the transition is proportional to the rate; that is, the transition probability per unit time;

$$I_{n_1 \to n_2} \propto \frac{dP_{n_2}(t)}{dt} \tag{5.73}$$

Eq.(5.73) demonstrates that the intensity is proportional to the square of the absolute value of the transition integral,

$$I_{n_1 \to n_2} \propto | < \psi_{n_2}|x|\psi_{n_1} > |^2 \tag{5.74}$$

The choice of x as the direction of the radiation field was arbitrary, in general all three transition integrals (called electric–dipole transition integrals)

$$< \psi_{n_2}|x|\psi_{n_1} > \quad , \quad < \psi_{n_2}|y|\psi_{n_1} > \quad , \quad < \psi_{n_2}|z|\psi_{n_1} > \tag{5.75}$$

must be considered. If one or more of the integrals of Eq.(5.75) is nonzero for a given transition, the corresponding line appears in the spectrum and the transition is said to be **allowed**. Transitions which are not allowed by the electric–dipole interactions are said to be **forbidden**.

Evaluation of these integrals yields the selection rules stated in Eq.(5.70).

One can write the bound state wave function for a one–electron atom in general

$$\psi_{n\ell m_\ell}(r,\theta,\phi) = R_{n\ell}(r)Y_{\ell m_\ell}(\theta,\phi) = R_{n\ell}(r)\Theta_{\ell m_\ell}(\theta)\Phi_{m_\ell}(\phi) \qquad (5.76)$$

where the radial wave function $R_{n\ell}(r)$ is in the form

$$R_{n\ell}(r) = -\left\{ \left(\frac{2Z}{na_\mu}\right)^3 \frac{(n-\ell-1)!}{2n[(n+\ell)!]^3} \right\}^{1/2} e^{-\rho/2}\rho^\ell L_{n+\ell}^{2\ell+1}(\rho) \qquad (5.77)$$

$\rho = (2Z/na_\mu)r$, $a_\mu = \hbar^2/e^2\mu$, μ is reduced mass. The explicit expression for the Laguerre polynomials $L_{n+\ell}^{2\ell+1}(\rho)$ is given by

$$L_{n+\ell}^{2\ell+1}(\rho) = \sum_{k=0}^{n_r}(-1)^{k+1}\frac{[(n+\ell)!]^2\rho^k}{(n_r-k)!(2\ell+1+k)!k!} \quad , \quad n_r = n-\ell-1 \quad (5.78)$$

The spherical harmonics $Y_{\ell m_\ell}(\theta,\phi)$ are given by

$$Y_{\ell m_\ell}(\theta,\phi) = (-1)^{m_\ell}\left[\frac{(2\ell+1)(\ell-m_\ell)!}{4\pi(\ell+m_\ell)!}\right]^{1/2} P_\ell^{m_\ell}(\cos\theta)e^{-im_\ell\phi} \quad , \quad m_\ell \geq 0$$
$$(5.79)$$

with the properties

$$Y_{\ell,-m_\ell}(\theta,\phi) = (-1)^{m_\ell}Y_{\ell m_\ell}^*(\theta,\phi) \quad , \quad |Y_{\ell m_\ell}(\theta,\phi)|^2 = \frac{1}{2\pi}|\Theta_{\ell m_\ell}(\theta)|^2 \quad (5.80)$$

θ dependent function $\Theta_{\ell m_\ell}(\theta)$ can be expressed in terms of the associated Legendre functions $P_\ell^{m_\ell}$ by

$$\Theta_{\ell m_\ell}(\theta) = (-1)^{m_\ell}\left[\frac{(2\ell+1)(\ell-m_\ell)!}{4\pi(\ell+m_\ell)!}\right]^{1/2} P_\ell^{m_\ell}(\cos\theta) \quad , \quad m_\ell \geq 0 \quad (5.81)$$

$$= (-1)^{|m_\ell|}\Theta_{\ell|m_\ell|}(\theta) \quad , \quad m_\ell < 0$$

Using these orthonormal wave functions one can verify the selection rules $\Delta\ell = \ell_1 - \ell_2 = \pm 1$ and $\Delta m_\ell = m_{\ell_1} - m_{\ell_2} = 0, \pm 1$ in the electric dipole approximation, namely considering the integrals $< \psi_{n_1\ell_1 m_{\ell_1}}|\alpha|\psi_{n_2\ell_2 m_{\ell_2}} >$,

$\alpha = x, y, z$, as follows:

Consider the following electric dipole integrals:

$$I =< \psi_{n'\ell'm'_\ell}|\alpha|\psi_{n\ell m_\ell} >=< n'\ell'm'_\ell|\alpha|n\ell m_\ell > \quad ; \quad \alpha = x, y, z \qquad (5.82)$$

a) Let $\alpha = z = r\cos\theta$

$$I = N \int_0^\infty R^*_{n'\ell'} r R_{n\ell} r^2 dr \qquad (5.83)$$

$$\times \int_0^\pi P^{m'_\ell}_{\ell'} \cos\theta P^{m_\ell}_\ell \sin\theta d\theta \int_0^{2\pi} e^{im'_\ell\phi} e^{-im_\ell\phi} d\phi$$

from the ϕ–dependence

$$< n'\ell'm'_\ell|z|n\ell m_\ell >= 0 \quad unless \quad m'_\ell = m_\ell \qquad (5.84)$$

b) Let

$$\alpha = x = r\sin\theta\cos\phi \qquad (5.85)$$

omitting the r and θ integrals

$$< n'\ell'm'_\ell|x|n\ell m_\ell >\propto \int_0^{2\pi} e^{im'_\ell\phi} \cos\phi e^{-im_\ell\phi} d\phi \qquad (5.86)$$

$$= \frac{1}{2} \int_0^{2\pi} \left\{ e^{i(m'_\ell - m_\ell + 1)\phi} + e^{i(m'_\ell - m_\ell - 1)\phi} \right\} d\phi \qquad (5.87)$$

$$= 0 \quad unless \quad m'_\ell = m_\ell \pm 1 \qquad (5.88)$$

with a similar expression for the y–component.

$\Delta m_\ell = 0$ for the electric vector $\parallel$ to the magnetic field (π polarization);

$\Delta m_\ell = \pm 1$ for the electric vector $\perp$ to the magnetic field (σ polarization).

The selection rules for ℓ follow from the θ–dependence. For the z component

$$I =< n'\ell'm'_\ell|z|n\ell m_\ell >\propto \int_0 P^{m'_\ell}_{\ell'} \cos\theta P^{m_\ell}_\ell \sin\theta d\theta \qquad (5.89)$$

using the relation

$$\cos\theta P_\ell^{m_\ell} = \frac{(\ell - m_\ell + 1)P_{\ell+1}^{m_\ell} + (\ell + m_\ell)P_{\ell-1}^{m_\ell}}{2\ell + 1} \tag{5.90}$$

and considering the orthogonality property of $P_\ell^{m_\ell}$

$$\int P_{\ell'}^{m_\ell'} \cos\theta P_\ell^{m_\ell} \sin\theta d\theta = 0 \quad unless \quad \ell' = \ell \pm 1 \tag{5.91}$$

Similarly, for x and y components we use

$$\sin\theta P_\ell^{m_\ell - 1} = \frac{P_{\ell+1}^{m_\ell} - P_{\ell-1}^{m_\ell}}{2\ell + 1} \tag{5.92}$$

to obtain the same rule $\ell' = \ell \pm 1$. Thus for electric dipole radiation $\Delta\ell = \pm 1$.

5.12   Angular momentum of bound electrons

The magnitude of the square of the electronic angular momentum vector in the hydrogen atom is given by

$$L^2 = \ell(\ell + 1)\hbar^2 \quad , \quad \ell = 0, 1, 2, ..., (n - 1) \tag{5.93}$$

and the component of the angular momentum along the z–axis is

$$L_z = m_\ell \hbar \quad , \quad m_\ell = 0, \pm 1, \pm 2, ..., \pm \ell \tag{5.94}$$

This directional limitation upon the angular momentum is sometimes called **space quantization**. This implies that the component of the angular momentum in a certain direction is quantised so that it can only take certain values. Indeed, the Bohr quantisation of angular momentum suggests that orbital angular momentum only occurs in integral units of $\hbar$.

To demonstrate the existence of space quantization, we place the atom in a magnetic field and observe the resulting changes in the energy levels.

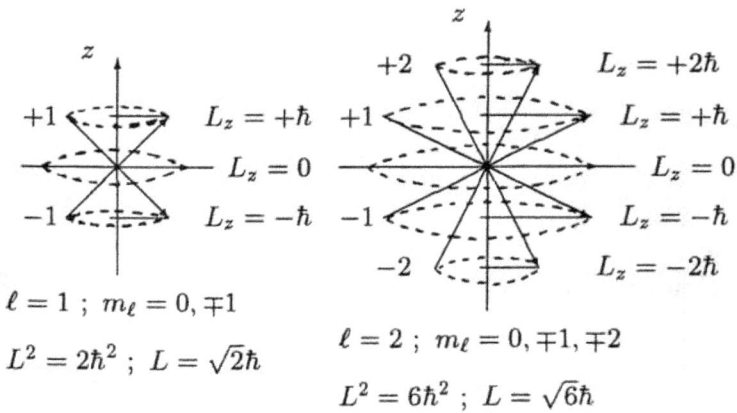

Fig. 5.7. Space quantization for $\ell = 1, 2$.

A moving particle with a charge q and a vector angular momentum **L** has a vector **magnetic dipole moment** μ_{mag},

$$\mu_{mag} = \gamma \mathbf{L} \tag{5.95}$$

The constant γ, called the **magnetogyric ratio**, is given by

$$\gamma = \frac{q}{2m_q c} \quad \left[= -\frac{e}{2m_e c} \quad (for \; electron) \right] \tag{5.96}$$

If a magnetic field is present, μ_{mag} interacts with the field **B**. Choosing the coordinate system so that **B** is in the z–direction, the **magnetic energy** is ($U = -\mu \cdot \mathbf{B}$)

$$\Delta E_{mag} = -(\mu_{mag})_z B = -\gamma L_z B \tag{5.97}$$

For a hydrogen atom in a magnetic field,

$$\Delta E_{mag} = -\left(\frac{-e}{2m_e c} \right) m_\ell \hbar B = m_\ell \left(\frac{e\hbar}{2m_e c} \right) B = m_\ell \mu_B B \tag{5.98}$$

The constant μ_B is called the Bohr magneton and has the value

$$\mu_B = \frac{e\hbar}{2m_e c} = 9.2732 \times 10^{-21} \; erg/G \tag{5.99}$$

According to Eq.(5.98), the presence of a magnetic field removes the degeneracy with respect to the quantum number m_ℓ. For example, if the hydrogen atom is in a 2p state ($\ell = 1$), the possible energy levels are ($E = E_0 + \Delta E_{mag}$):

$$E(2p_0) = -\frac{1}{4}\frac{e^2}{2a_0} \quad , \quad E(2p_{\mp1}) = -\frac{1}{4}\frac{e^2}{2a_0} \mp \mu_B B \qquad (5.100)$$

Bohr magneton : μ_B

In cgs unit system: $[\mu_B] \equiv (erg/gauss)$

$$\mu_B = \frac{e\hbar}{2m_e c} = \frac{4.80292 \times 10^{-10}(esu) \times 1.0545 \times 10^{-27}(erg \cdot s)}{2 \times 9.109069 \times 10^{-28}(g) \times 2.997925 \times 10^{10}(cm/s)}$$

$$= \frac{4.80292 \times 1.0545 \times 10^{-37}}{2 \times 9.109069 \times 2.997925 \times 10^{-18}} \quad erg/gauss$$

$$= 0.09273 \times 10^{-19} \quad erg/gauss = 9.273 \times 10^{-21} \quad erg/gauss$$

In SI unit system: $[\mu_B] \equiv (J/T)$

$$\mu_B = \frac{e\hbar}{2m_e} = \frac{1.60208 \times 10^{-19}(C) \times 1.0545 \times 10^{-34}(J \cdot s)}{2 \times 9.109069 \times 10^{-31}(kg)}$$

$$= \frac{1.60208 \times 1.0545 \times 10^{-53}}{2 \times 9.109069 \times \times 10^{-31}} \quad J/T$$

$$= 0.09273 \times 10^{-22} \quad J/T = 9.273 \times 10^{-24} \quad J/T$$

$1 \ erg/gauss \equiv 1\frac{10^{-7}J}{10^{-4}T} = 10^{-3} \ J/T$.

If the atom is in an s state, $m_\ell = 0$, and no change of energy is expected in a magnetic field. We can therefore test Eq.(5.63) by observing the $2p \leftrightarrow 1s$ transition in the presence of a magnetic field. The selection rules for the possible transitions, Eq.(5.52), show that $2p_0 \leftrightarrow 1s$, $2p_{+1} \leftrightarrow 1s$, $2p_{-1} \leftrightarrow 1s$ are all allowed, so that three equally spaced lines are expected. It is referred to as a **Zeeman triplet**. The limiting Zeeman behavior of the lines in high magnetic fields is called the **Paschen–Back effect**.

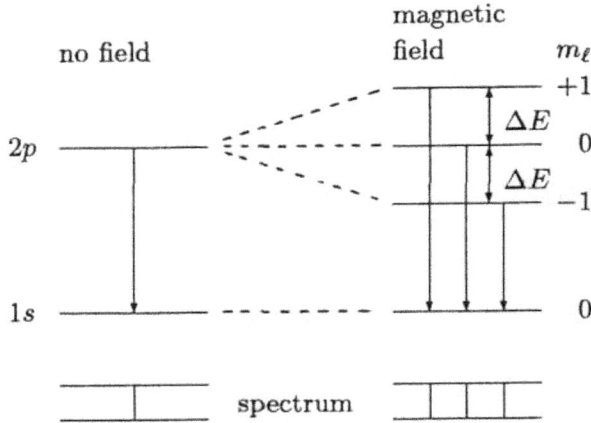

For $B = 5 \times 10^4\ G$, $\Delta E \cong 2.9 \times 10^{-4}\ eV$

Fig. 5.8. Zeeman triplet for $2p \rightarrow 1s$ transition.

5.13 Spin of electrons

When the Zeeman triplet is examined under high resolution, the two outer lines are found to be split into two lines each, giving five lines in all.

Observations of corresponding anomalies in the spectra of many-electron atoms suggested to Goudsmit and Uhlenbeck in 1925 that the electron has an intrinsic magnetic moment independent of its orbital motion.

If we associate this magnetic moment with an intrinsic angular momentum or **spin**, we can treat it analogously to the orbital angular momentum. We associate the quantum number s with the magnitude of the total spin angular momentum S and the quantum number m_s

with its z component S_z, such that

$$S^2 = s(s+1)\hbar^2 \tag{5.101}$$

and

$$S_z = m_s\hbar \quad ; \quad m_s = -s, -s+1, ..., +s \tag{5.102}$$

The quantum numbers s and m_s are the spin analogs to ℓ and m_ℓ for orbital angular momentum. The spin quantum number s has been found to have only a single value, $s = 1/2$. Therefore, $S^2 = 3\hbar^2/4$. m_s can have the values $\pm 1/2$, $S_z = \pm\hbar/2$. Thus, there are only two possible orientations of the spin angular momentum.

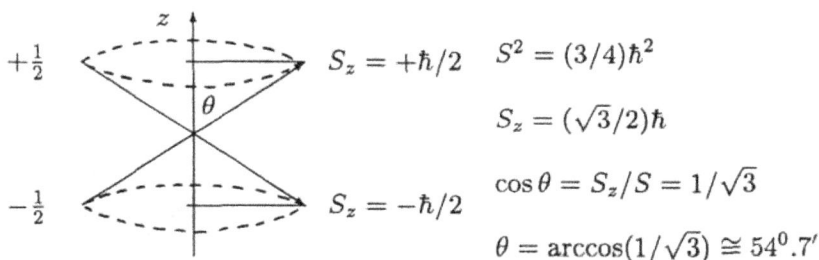

$$S_z = +\hbar/2 \quad S^2 = (3/4)\hbar^2$$
$$S_z = (\sqrt{3}/2)\hbar$$
$$S_z = -\hbar/2$$
$$\cos\theta = S_z/S = 1/\sqrt{3}$$
$$\theta = \arccos(1/\sqrt{3}) \cong 54^0.7'$$

Fig. 5.9. Spin quantization for a single electron.

The z component of the intrinsic magnatic moment is

$$(\mu_s)_z = \gamma_s S_z = m_s \gamma_s \hbar \tag{5.103}$$

where γ_s is the gyromagnetic ratio for the spin angular momentum.

Examination of the splitting of spectral lines in magnetic fields has shown that the value of γ_s is almost exactly twice the gyromagnetic ratio for the orbital motion of the electron, that is

$$\gamma_s = 2\gamma_\ell = -\frac{e}{m_e c} \tag{5.104}$$

The magnetic energy of a one–electron atom in a very strong magnetic field due to the presence of electron spin takes the form (for field direction along the z–axis)

$$\Delta E_{mag} = m_\ell \left(\frac{e\hbar}{2m_e c}\right) B + m_s \left(\frac{e\hbar}{m_e c}\right) B$$

$$= \frac{e\hbar}{2m_e c} B(m_\ell + 2m_s) = \mu_B B(m_\ell + 2m_s) \qquad (5.105)$$

If we consider $2p$ and $1s$ states again, we have for $2p$, $m_\ell = 0, \pm 1$, $m_s = \pm 1/2$, and for $1s$, $m_\ell = 0$, $m_s = \pm 1/2$. The resulting energy levels and spectrum look like:

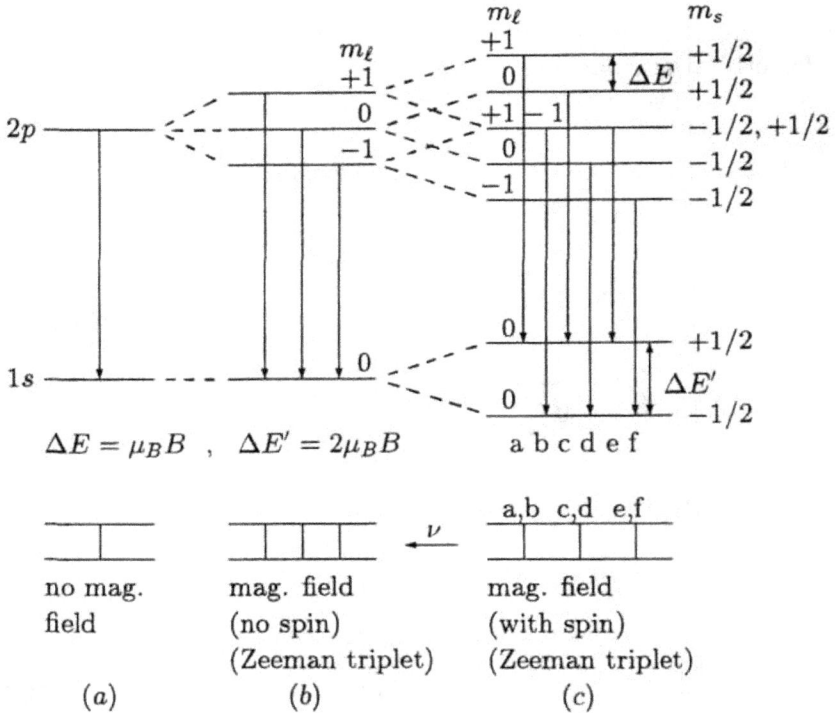

Fig. 5.10. Splitting of energy levels in magnetic field.

Each level in (b) is now split into two levels, corresponding to the two possible spin orientations.

For the allowed transitions the selection rule for the quantum number m_s in the case of electric dipole induced transitions is

$$\Delta m_s = 0 \qquad (5.106)$$

that is, the electron–spin orientation does not change. We find six allowed transitions as shown in figure (c) above, but only three distinct frequencies are expected from the six transitions, these are

$$\tilde{\nu} = \frac{E_{2p} - E_{1s}}{hc} + \frac{\mu_B}{hc} B \qquad \Delta m_\ell = +1 \quad (a, b)$$

$$\tilde{\nu} = \frac{E_{2p} - E_{1s}}{hc} \qquad \Delta m_\ell = 0 \quad (c, d) \qquad (5.107)$$

$$\tilde{\nu} = \frac{E_{2p} - E_{1s}}{hc} - \frac{\mu_B}{hc} B \qquad \Delta m_\ell = -1 \quad (e, f)$$

Thus, from the one–electron theory including the electron spin, the results appear to be a normal Zeeman triplet identical to that obtained previously without spin.

Still there is a quantitative disagreement with experiment, because the two outer lines [in fig.(c), and Eq.(5.107)] are both resolvable into two lines.

The Stern–Gerlach Experiment :

The existence of electron spin and of the magnitude of the associated magnetic moment was given by O. Stern and W. Gerlach (in 1922) in their experiments with beams of alkali metal atoms (Li, Na, K, Rb, Cs).

An alkali metal atom has a single electron in an s orbital whose intrinsic spin angular momentum makes the dominant contribution to the atomic magnetic moment. All the other electrons occupy orbitals in such a way that their orbital and spin magnetic moments cancel.

If a beam of atoms is placed in a magnetic field directed along the z–axis, the atoms acquire additional potential energy.

$$\Delta E_{mag} = -(\mu_{mag})_z B \qquad (5.108)$$

The magnetic moment is defined as $(\mu_{mag})_z = m_s \gamma_s \hbar$. If the external magnetic field is uniform, the atoms do not experience any magnetic force. However, if the external magnetic field is inhomogeneous, say $B = B_0 + \alpha z$; an atom traveling in the y direction in the inhomogeneous field has a magnetic potential energy

$$\Delta E_{mag} = -(\mu_{mag})_z B_0 + \alpha z) \qquad (5.109)$$

The magnetic force experienced by the atom is

$$F_{mag} = -\frac{d\Delta E_{mag}}{dz} = (\mu_{mag})_z \alpha \qquad (5.110)$$

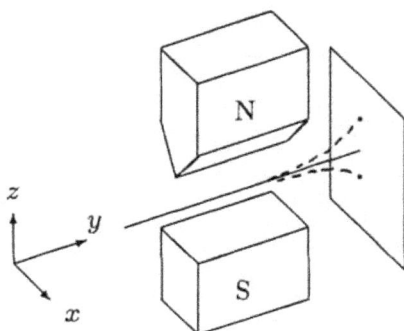

Fig. 5.11. Schematic diagram of the Stern–Gerlach experiment.

The direction of force is in the positive or negative x direction, depending upon the sign of $(\mu_{mag})_z$.

The Stern–Gerlach experiment shows that an alkali atom beam is split into two beams of equal intensity, indicating that there are only two possible values of the intrinsic angular momentum.

Furthermore, the beams are deflected an equal distance from the y-axis, one in the positive z direction, and one in the negative z direction. This indicates that the two values of $(\mu_{mag})_z$ are of equal magnitude and opposite sign. Thus, the theory of electron spin is directly confirmed by the Stern–Gerlach experiment.

5.14   Coupling of states

Often we work with systems made up of two or more parts, each with angular momenta. These may be different particles, or perhaps the spin and orbital properties of one particle.

Coupling of two angular momenta:

Consider a system in which the total angular momentum $\mathbf{J}$ is the sum of components $\mathbf{J_1}$ and $\mathbf{J_2}$. If an interaction between the two parts is such as to leave the individual angular momenta and their z–components constants of the motion, a complete set of commuting operators would include H, J_1^2, J_{1z}, J_2^2, J_{2z}. The corresponding eigenfunctions $|j_1 j_2 m_1 m_2 >$ may always be written in the simple product form

$$|j_1 j_2 m_1 m_2 > = |j_1 m_1 > |j_2 m_2 > \qquad (5.111)$$

We have the eigenvalue equations

$$J_1^2 |j_1 j_2 m_1 m_2 > = j_1(j(1+1)|j_1 j_2 m_1 m_2 > \qquad (5.112)$$

$$J_{1z} |j_1 j_2 m_1 m_2 > = m_1 |j_1 j_2 m_1 m_2 > \qquad (5.113)$$

for J_1^2 and J_{1z} and similarly for J_2^2 and J_{2z}. We could, however, choose a set including H, J_1^2, J_2^2, $J^2 = (\mathbf{J_1} + \mathbf{J_2})^2$ and $J_z = J_{1z} + J_{2z}$, which contains as many physical observables as before. The eigenfunctions $|j_1 j_2 J M >$ now satisfy

$$J^2 |j_1 j_2 J M > = J(J+1)|j_1 j_2 J M > \qquad (5.114)$$

$$J_z |j_1 j_2 J M > = M |j_1 j_2 J M > \qquad (5.115)$$

while J_1^2 and J_2^2 have the same eigenvalues as before. In physical applications this is often a more useful set.

For instance when an interaction between the two parts of the system is introduced as a perturbation J^2 and J_z may be conserved, but not the individual z–components J_{1z} and J_{2z}.

The unitary transformation connecting these two representations defines the **vector–addition coefficients**:

$$|j_1 j_2 J M > = \sum_{m_1 m_2} |j_1 j_2 m_1 m_2 > < j_1 j_2 m_1 m_2 |j_1 j_2 J M > \qquad (5.116)$$

$$|j_1 j_2 m_1 m_2 > = \sum_{J M} |j_1 j_2 J M > < j_1 j_2 J M |j_1 j_2 m_1 m_2 > \qquad (5.117)$$

$$< j_1 j_2 m_1 m_2 | j_1 j_2 J M > = < j_1 j_2 J M | j_1 j_2 m_1 m_2 > \qquad (5.118)$$

sometimes these coupling coefficients are called as **Wigner** or **Clebsch–Gordon** coefficients. For short–hand notation they are written as $< j_1 j_2 m_1 m_2 | J M >$.

For given j_1 and j_2 the values of J are restricted by the **triangular condition**

$$j_1 + j_2 \geq J \geq |j_1 - j_2| \qquad (5.119)$$

and J ranges from $j_1 + j_2$ down to $|j_1 - j_2|$ in integer steps.

Since $J_z = J_{1z} + J_{2z}$, the vector addition coefficients vanishes unless $M = m_1 + m_2$.

The orthonormality of the eigenfunctions $|JM>$ and $|j_1 j_2 m_1 m_2 >$ leads to the orthogonality relations for the coefficients

$$\sum_{m_1 m_2} < J M | j_1 j_2 m_1 m_2 > < j_1 j_2 m_1 m_2 | J' M' > = \delta_{JJ'} \delta_{MM'} \qquad (5.120)$$

and

$$\sum_{JM} < j_1 j_2 m_1 m_2 | J M > < J M | j_1 j_2 m_1' m_2' > = \delta_{m_1 m_1'} \delta_{m_2 m_2'} \qquad (5.121)$$

which express the unitary nature of the transformations.

Since each coefficient vanishes unless $M = m_1 + m_2$, the sum over M is purely formal in the second orthogonality relation and in fact the sum is only over J. From a dynamical point of view the transformations given in Eqs.(5.116,117) describe the addition of angular momentum. The geometric interpretation is as following:

The state $|j_1 j_2 J M >$ is represented by the two vectors $\mathbf{J}_1$ and $\mathbf{J}_2$ precessing in phase about their resultant $\mathbf{J}$ (which in turn precesses about the z–axis). The precession of $\mathbf{J}_1$ and $\mathbf{J}_2$ about $\mathbf{J}$ and its projection on the z–axis represents the indeterminicy in their individual z–components m_1 and m_2 although their sum M remains constant.

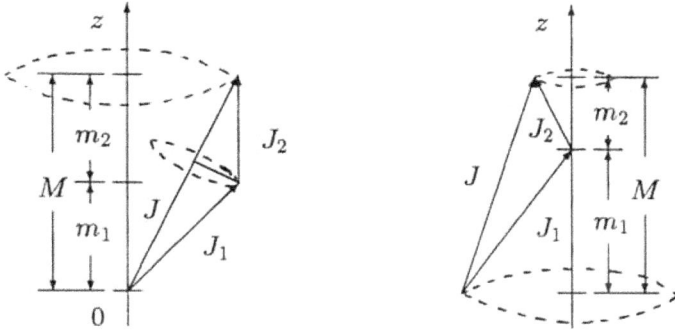

Fig. 5.12. Coupling models.

The square of the vector addition coefficient

$$| < j_1 j_2 m_1 m_2 | JM > |^2 \qquad (5.122)$$

is the probability that a measurement in the state $|JM >$ gives the particular values m_1 and m_2 for J_{1z} and J_{2z} . Conversly in the state $|j_1 j_2 m_1 m_2 >$ the two vectors precess independently about the z–axis and

$$| < JM | j_1 j_2 m_1 m_2 > |^2 \qquad (5.123)$$

is the probability that at any instant their resultant will be J.

Recurrence relations for the coefficients can be obtained from the operator identities

$$J_\pm = J_{1_\pm} + J_{2_\pm} \qquad (5.124)$$

In matrix form these become

$$< JM \pm 1 | J_\pm | JM > =$$

$$\sum_{m_1 m_2 n_1 n_2} < JM \pm 1 | n_1 n_2 > < n_1 n_2 | J_{1_\pm} + J_{2_\pm} | m_1 m_2 > < m_1 m_2 | JM >$$

$$(5.125)$$

Matrix multiplication from the left by $< m_1' m_2' | JM' >$, use of the orthogonality relation equation, and the matrix elements of J_z and $J_\mp$

$$< jm | J_z | jm > = m \qquad (5.126)$$

and

$$< jm \mp 1|J_\pm|jm> = [(j \pm m + 1)(j \mp m)]^{1/2} \tag{5.127}$$

one obtaines

$$[(J \pm M + 1)(J \mp M)]^{1/2} < m_1 m_2|JM \pm 1> = \tag{5.128}$$

$$[(j_1 \mp m_1 + 1)(j_1 \pm m_1)]^{1/2} < m_1 \mp 1, m_2|JM>$$

$$+ [(j_2 \mp m_2 + 1)(j_2 \pm m_2)]^{1/2} < m_1 m_2 \mp 1|JM>$$

These relations are sufficient to determine the vector addition coefficients.

Racah obtained the general formula:

$$< j_1 j_2 m_1 m_2|JM> = \delta_{m_1+m_2,M} \, (\Delta_1 \times \Delta_2)^{1/2} \times \sum_\nu \frac{(-1)^\nu}{\Delta_3} \tag{5.129}$$

where

$$\Delta_1 = \frac{(j_1 + j_2 - J)!(j_1 + J - j_2)!(j_2 + J - j_1)!}{j_1 + j_2 + J + 1)!} \tag{5.130}$$

$$\Delta_2 = (2J + 1)(j_1 + m_1)!(j_1 - m_1)!(j_2 + m_2)!(j_2 - m_2)! \tag{5.131}$$
$$\times (J + M)!(J - M)!$$

$$\Delta_3 = \nu!(j_1 - m_1 - \nu)!(J - j_2 + m_1 + \nu)!(j_2 + m_2 - \nu)! \tag{5.132}$$
$$\times (J - j_1 - m_2 + \nu)!(j_1 + j_2 - J - \nu)!$$

and ν runs over all values which do not led to negative factorials.

5.15 Term symbols for one–electron atoms

Since there is only one occupied orbital in the case of one–electron atoms and ions, the quantum numbers that identify the orbital occupied by the electron also identify the electronic state of the atom as a whole. Thus, the atomic energy levels (called **terms** in spectroscopic language) are designated by symbols which differ from those used for orbitals. In atomic term symbols, the capital letters $S, P, D, F, G, \ldots$ are employed for the total electronic orbital angular momentum quantum numbers, $L = 0, 1, 2, 3, 4, ...$, respectively. For one–electron atom or ions the term symbols $S, P, D, F, G, ...$ correspond exactly to the orbital designations $s, p, d, f, g, ...$

A left superscript to the L symbol gives the spin degeneracy $2S + 1$ of the atom which is $2s + 1 = 2$ for a one–electron system, and a right subscript gives the value of the quantum number $J = L + S, L + S - 1$, ..., $|L - S|$ for the total angular momentum, $\mathbf{J} = \mathbf{L} + \mathbf{S}$. The principal quantum number n of the occupied orbital is sometimes written to the left of the term symbol. Thus, the term symbol has the form

$$n \left({}^{2S+1}\mathrm{L}_J \right) \tag{5.133}$$

With these definitions, the ground state of the hydrogen atom ($n = 1$, $L = \ell = 0$, $S = s = 1/2$, and $J = 1/2$) has the term symbol $1({}^2S_{1/2})$. The two $2p$ states ($n = 2$, $L = 1$, $S = 1/2$, $J = L \pm S$) have the term symbols $2({}^2P_{3/2})$ and $2({}^2P_{1/2})$. Terms such as these with the same values of n, L, and S, but with different values of J, differ slightly in energy because of spin–orbit coupling.

5.16 Spin–orbit coupling

The additional splitting can be understood in the framework of the relativistic treatment of the hydrogen atom, given by P.A.M. Dirac in 1928. One of the correction terms is due to the spin–orbit interaction, called **spin–orbit coupling**, which arises from the interaction between the magnetic moments associated with the orbital and spin angular momenta of the electron. The magnetic dipole moment due to the orbital motion creates a magnetic field which interacts with the electron spin magnetic moment.

If an atom is in a very strong magnetic field, the effect of the spin–orbit coupling is to shift all the levels with $m_\ell m_s > 0$ upward slightly and those with $m_\ell m_s < 0$ downward slightly.

Also, the remaining degeneracy in the $2p$ state is lifted. The transitions are correspondingly shifted. Thus, transitions (a,b) now have slightly different frequencies; also (e,f). Therefore, the so–called normal Zeeman pattern of three lines is seen to consist of five lines (**anomalous Zeeman effect**). This is in agreement with experimental observations. As the external magnetic field is decreased in magnitude and the spin–orbit coupling becomes relatively more important, changes in the spectrum occur. Also, there are some very small additional effects, e.g. the Lamb shift (discovered in 1947), that have been observed by refined spectroscopic techniques and require a relativistic quantum–mechanical treatment of the interaction between radiation and the atom for their explanation.

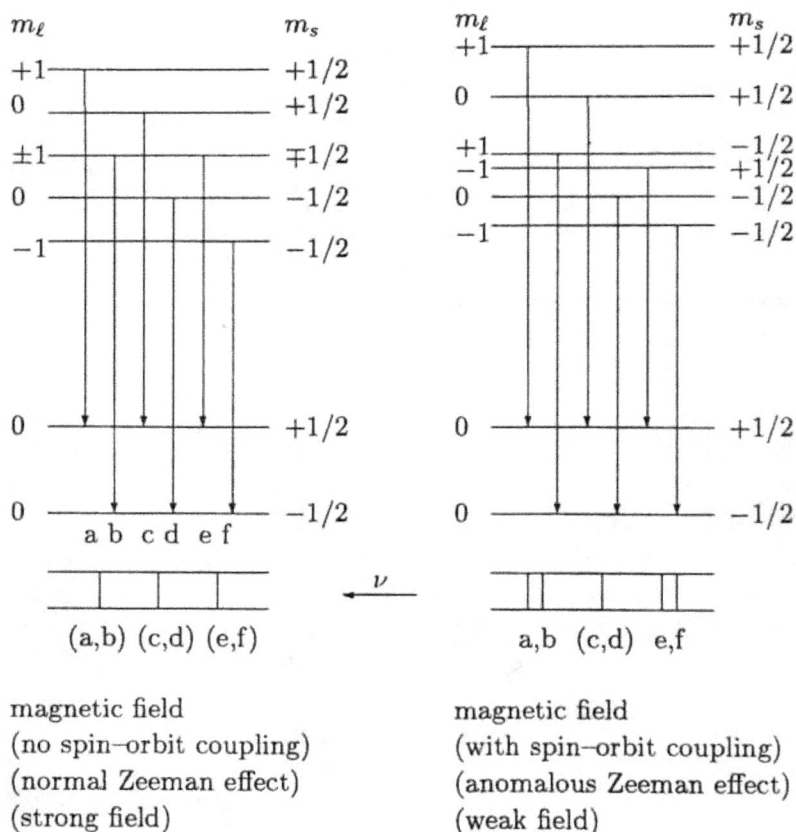

magnetic field
(no spin–orbit coupling)
(normal Zeeman effect)
(strong field)

magnetic field
(with spin–orbit coupling)
(anomalous Zeeman effect)
(weak field)

Fig. 5.13. Normal and anomalous Zeeman effects.

5.17   Some remarks on the Zeeman effect

In 1896, P. Zeeman observed that the spectral lines of atoms were split in the presence of an external magnetic field. The reason of splitting of spectral lines in magnetic field is due to the interaction of external magnetic field and the magnetic dipole moment of the atom.

The Schrödinger equation for a one–electron atom in a constant magnetic field, including the spin–orbit interaction, is

$$\left[-\frac{\hbar^2}{2m}\nabla^2 - \frac{Ze^2}{r} + \xi(r)\mathbf{L}\cdot\mathbf{S} + \frac{\mu_B}{\hbar}(\mathbf{L}+2\mathbf{S})\cdot\mathbf{B}\right]\psi(r) = E\psi(r)$$

$$(5.134)$$

For a strong magnetic field the contribution of $\xi(r)\mathbf{L}\cdot\mathbf{S}$ term is negligible with respect to the contribution of $(\mu_B/\hbar)(\mathbf{L}+2\mathbf{S})\cdot\mathbf{B}$ term, for weak field the contributions are reversed.

$$\xi(r) = \left(\frac{1}{2m^2c^2}\right)\frac{1}{r}\frac{\partial V(r)}{\partial r} = \frac{Ze^2}{2m^2c^2r^3} \qquad (5.135)$$

$\xi(r)\mathbf{L}\cdot\mathbf{S}$: Spin–orbit coupling (anomalous Zeeman effect; weak field)
$(\mu_B/\hbar)(\mathbf{L}+2\mathbf{S})\cdot\mathbf{B}$: (normal Zeeman effect; strong field)

For strong field case the contribution of the interaction energy to the total energy is

$$\Delta E_{mag} = \mu_B B(m_\ell + 2m_s) \qquad (5.136)$$

The introduction of the magnetic field does not remove the degeneracy in ℓ, but by providing a preferred direction in space, it does remove the degeneracy in m_ℓ and m_s, splitting each level with a given n into equally spaced terms. For example, for the case of a p level ($\ell = 1$), see Fig. 5.14, the energy of the states with $m_\ell = +1$ and $m_s = -1/2$ coincides with those with $m_\ell = -1$ and $m_s = +1/2$. In the strong field limit we do not consider the spin–orbit coupling. The selection rules for electric dipole transitions require $\Delta m_s = 0$ and $\Delta m_\ell = 0, \pm 1$.

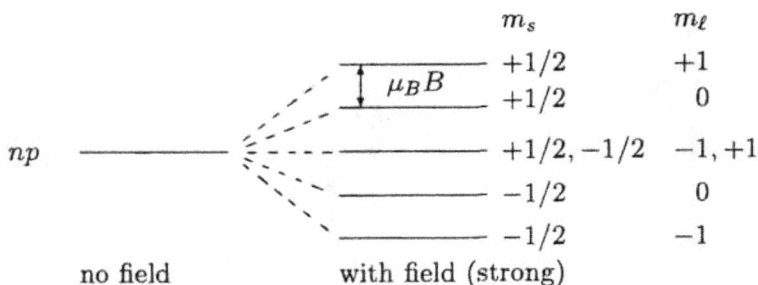

Fig. 5.14. Splitting of a p level in strong magnetic field.

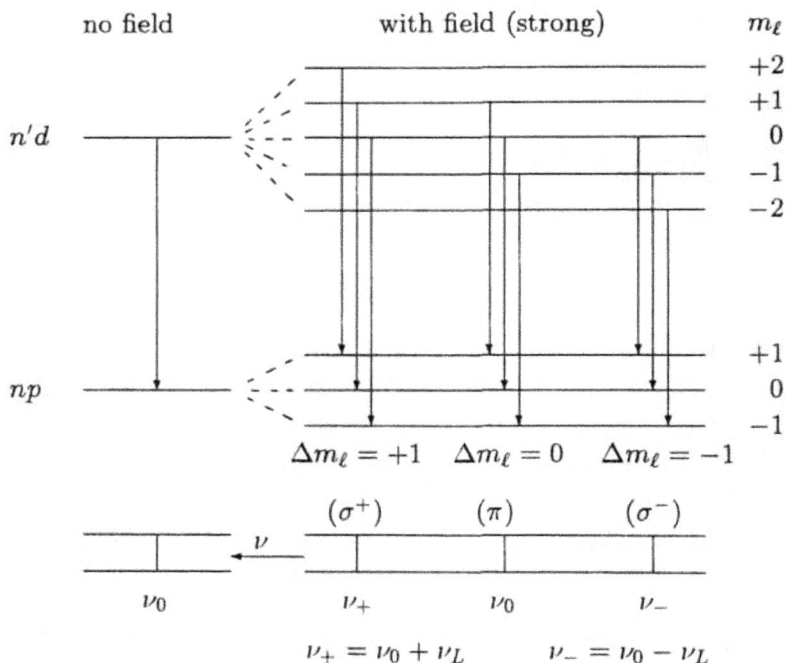

Fig. 5.15. Normal Zeeman effect (Zeeman triplet).

Thus the spectral line corresponding to a transition $n \rightarrow n'$ is split into **three** components. The line corresponding to $\Delta m_\ell = 0$ has the original frequency $\nu_{nn'}$ and is called the π line, while the two lines with

$\Delta m_\ell = \pm 1$ are called σ lines and correspond to frequencies

$$\nu_{nn'}^{\pm} = \nu_{nn'} \pm \nu_L \tag{5.137}$$

where

$$\nu_L = \frac{\mu_B}{h} B \tag{5.138}$$

is known as the **Larmor frequency**. This splitting is called the normal Zeeman effect, and three lines are said to form a **Lorentz triplet**. See Fig. 5.15 for $n'd \rightarrow np$ transition.

In a strong magnetic field nine transitions are possible between the split levels. There are only three different frequencies and the lines form a Lorentz triplet. The frequencies of transitions associated with $m_s = -1/2$ are the same as those for $m_s = +1/2$.

The Lorentz triplet can be observed in many–electron atoms for which the total spin is zero (singlet lines), as in this case the spin–orbit coupling vanishes.

For observations made in a direction perpendicular to the field, the lines with $\Delta m_\ell = 0$ are polarized parallel to the field (π–component), the lines with $\Delta m_\ell = \pm 1$, perpendicular to the field (σ–components).

For weak field case the Zeeman effect is known as the anomalous Zeeman effect. The energy of interaction in this case is calculated as

$$\Delta E_{mag}^{m_j} = g \mu_B B_z m_j \tag{5.139}$$

where g is called the **Lande's g factor** and is given by

$$g = 1 + \frac{j(j+1) + s(s+1) - \ell(\ell+1)}{2j(j+1)} \tag{5.140}$$

where j is the total angular momentum quantum number and m_j is the corresponding z–component which satisfy the relations

$$J^2 = j(j+1)\hbar^2 \quad and \quad J_z = m_j \hbar \;\; ; \quad m_j = j, j-1, ..., -j. \tag{5.141}$$

Polarization of radiation emitted in an electric dipole transition:

a) Longitudinal observation (π line absent)

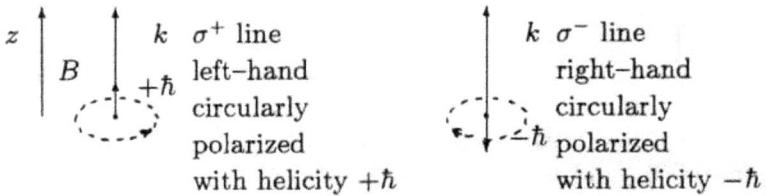

b) Transversal observation (the intensity of the π
component is twice that of each σ component)

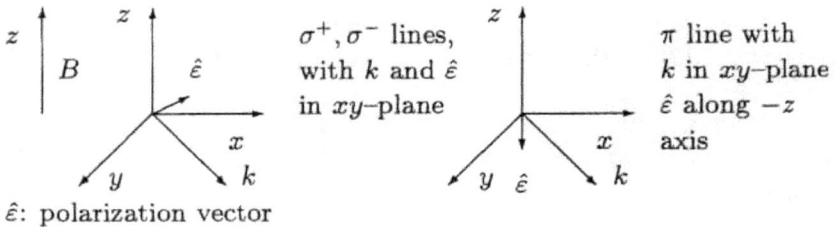

$\hat{\varepsilon}$: polarization vector

Fig. 5.16. Polarization of emitted radiation.

At field strengths for which the spin–orbit interaction is appreciable,
but still small compared with the term in **B**, the contribution of **L · S**
term to the total energy is obtained as

$$\Delta E = \lambda_{n\ell} m_\ell m_s \quad for \quad \ell \neq 0 \quad ; \quad \Delta E = 0 \quad for \quad \ell = 0 \qquad (5.142)$$

where

$$\lambda_{n\ell} = -\frac{\alpha^2 Z^2}{n} E_n \frac{1}{\ell(\ell + 1/2)(\ell + 1)} \quad , \quad \ell \neq 0 \qquad (5.143)$$

the degeneracy in ℓ is removed. The energy of the level $|n\ell m_\ell m_s >$ now
takes the form

$$E_{n\ell m_\ell m_s} = E_n + \mu_B B_z m_\ell + \lambda_{n\ell} m_\ell m_s \qquad (5.144)$$

The energy difference between the levels $n\ell m_\ell$ and $n'\ell'm'_\ell$ when $m_s = m'_s$
is

$$\delta E = E_{n'} - E_n + \mu_B B_z(m'_\ell - m_\ell) + (\lambda_{n'\ell'} m'_\ell - \lambda_{n\ell} m_\ell)m_s \qquad (5.145)$$

This expression gives the frequencies $\delta E/h$ of the observed lines, with $\Delta m_\ell = m'_\ell - m_\ell$ restricted to the values $0, \pm 1$. The observed splitting in this case is known as the **Paschen–Back effect**. In other words, with increasing field strength, when the magnetic splitting becomes greater than the multiplet splitting, F. Paschen and E. Back found in 1912 that the anomalous Zeeman effect changes over to the normal. See Fig. 5.17 for the splitting of $^2P_{3/2}$, $^2P_{1/2}$ and $^2S_{1/2}$ levels of H–atom in a weak magnetic field and possible electric dipole transitions.

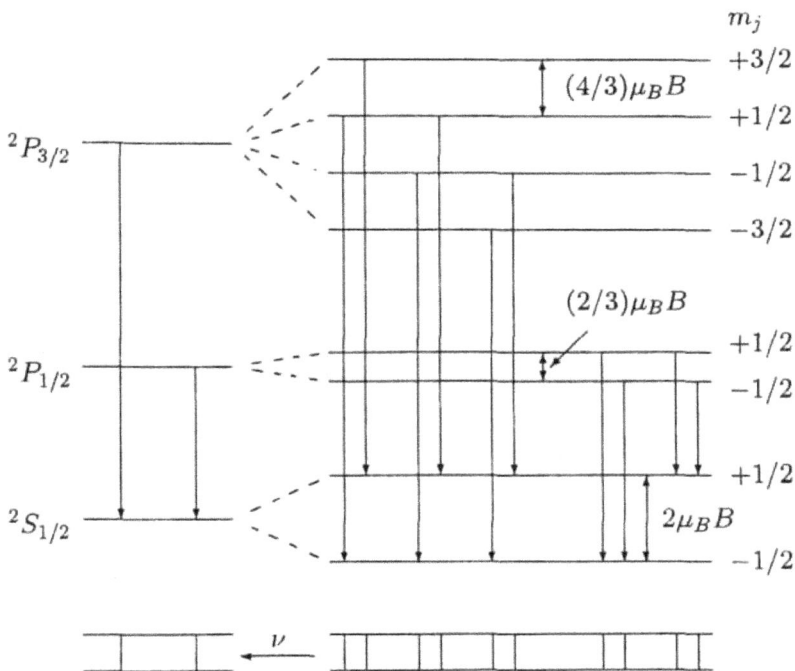

Fig. 5.17. Splitting of ^2P and ^2S levels of H–atom in a weak magnetic field and possible transitions.

Transitions with equal Δm_ℓ nearly coincide, whereas in the normal Zeeman effect they exactly coincide. Thus, each component of **normal** triplet has two components. For small B_z, the splitting is uneven (the anomalous Zeeman effect), but for large B_z, the splitting is even and only three lines are seen (Paschen– Back effect).

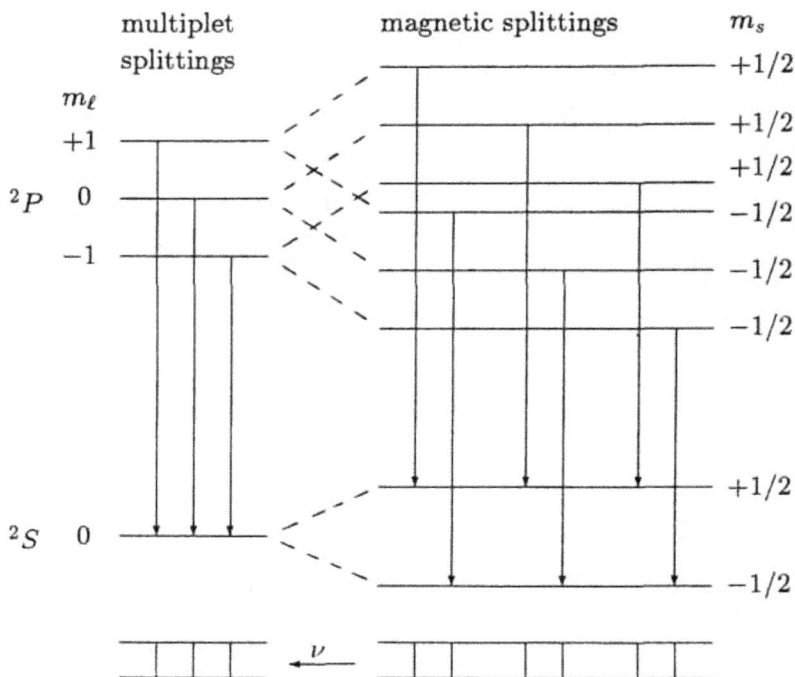

Fig. 5.18. Paschen–Back effect for $^2P \rightarrow ^2S$ transition.

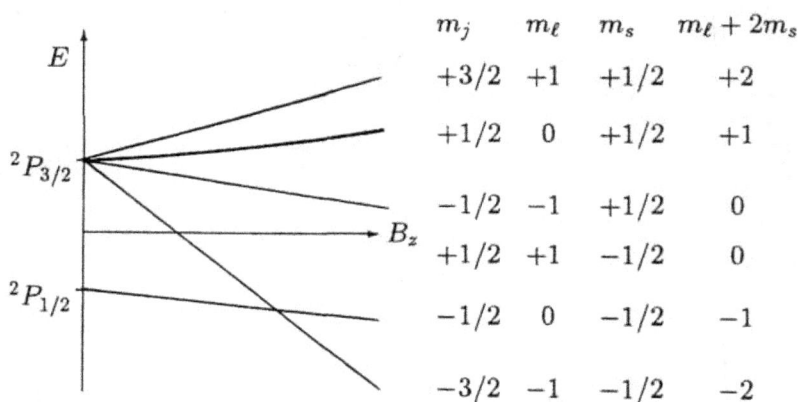

Fig. 5.19. Level splittings with respect to field strength.

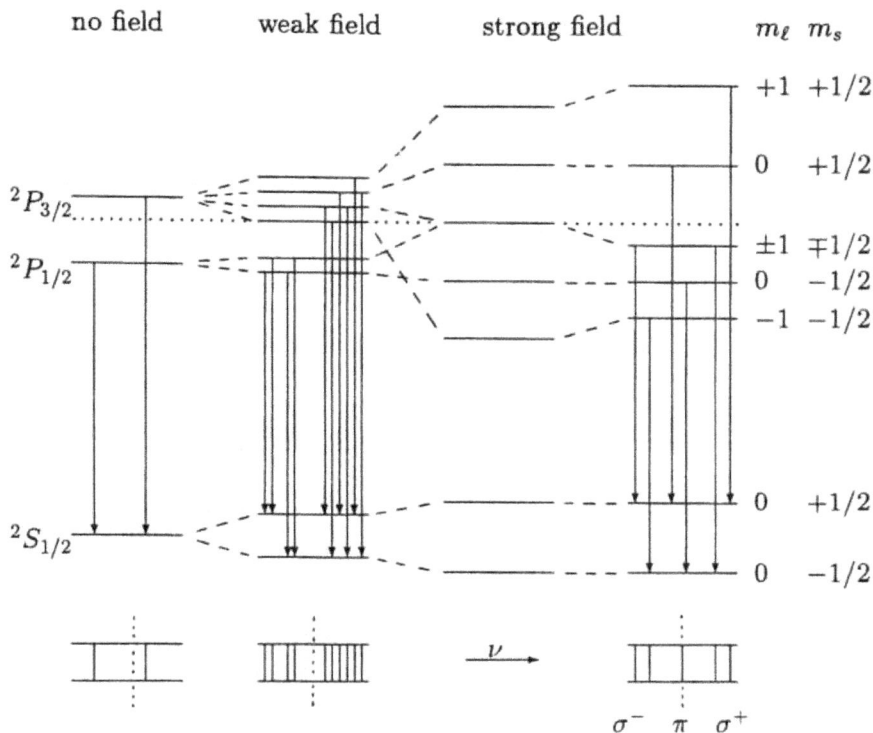

Fig. 5.20. Detailes of Paschen–Back effect for $^2P \rightarrow ^2S$ transitions.

5.18 The Stark effect

In the presence of an electric field $\vec{\mathcal{E}}$ the atomic levels will also be split. This effect was first observed by J. Stark in 1913. The interaction here is between $\vec{\mathcal{E}}$ and the induced electric dipole moment $\mathbf{d}$ as a result of electron polarization.

Let us consider the Stark effect on hydrogenlike atoms. Using time–independent perturbation theory we can calculate the energy of interaction, the effect of external electric field on the energy levels. The unperturbed Hamiltonian is

$$H_0 = -\frac{\hbar^2}{2m}\nabla^2 - \frac{Ze^2}{r} \tag{5.146}$$

whose eigenfunctions we denote by $\phi_{n\ell m_\ell}(\mathbf{r})$. If we take the electric field along z–direction, then the perturbing potential is

$$H_1 = e\mathbf{r} \cdot \vec{\mathcal{E}} = e\mathcal{E}z \tag{5.147}$$

where $\mathcal{E}$ is the electric field strength. The first order energy shift of the ground state, which is a nondegenerate, is given by the expression

$$E_{100}^{(1)} = e\mathcal{E} < \phi_{100}|z|\phi_{100} >= 0 \tag{5.148}$$

Thus for the ground state there is no energy shift that is linear in the electric field $\mathcal{E}$. Classically, a system that has an electric dipole moment $\mathbf{d}$, will experience an energy shift of magnitude $-\mathbf{d} \cdot \vec{\mathcal{E}}$. Thus an atom, in its ground state, has no permanent dipole moment. In general one can say that, systems in nondegenerate states cannot have permanent dipole moments.

Let us look at the second order perturbation term. It reads

$$E_{100}^{(2)} = e^2\mathcal{E}^2 \sum_{n\ell m_\ell} \frac{|< \phi_{n\ell m_\ell}|z|\phi_{100} >|^2}{E_1^{(0)} - E_n^{(0)}} \tag{5.149}$$

The matrix element here is

$$< \phi_{n\ell m_\ell}|z|\phi_{100} >= \int d^3 r\, R_{n\ell m_\ell}(r)Y_{\ell m_\ell}^*(\theta,\phi)r\cos\theta R_{100}(r)Y_{00}(\theta,\phi) \tag{5.150}$$

using $Y_{00} = 1/\sqrt{4\pi}$ and $\cos\theta = \sqrt{4\pi/3}Y_{10}$, the angular part of the integral becomes

$$\int d\Omega Y_{\ell m_\ell}^*(\theta,\phi)\frac{1}{\sqrt{3}}Y_{00}(\theta,\phi) = \frac{1}{\sqrt{3}}\delta_{\ell 1}\delta_{m_\ell 0} \tag{5.151}$$

The fact that the m_ℓ–value must be the same for the two states is one of the selection rules, $\Delta m_\ell = 0$. It follows from the fact that $[L_z, z] = 0$, the perturbation commutes with L_z. The radial part has the integral

$\int r^2 dr\, R_{n10}(r)\, r\, R_{100}(r)$. This integral can be evaluated. Therefore the result is

$$| < \phi_{n10}|z|\phi_{100} > |^2 = \frac{2^8 n^7 (n-1)^{2n-5}}{3(n+1)^{2n+5}} a_0^2 = f(n)a_0^2 \qquad (5.152)$$

For the second order shift in the ground state, this gives

$$E_{100}^2 = -e^2 \mathcal{E}^2 a_0^2 \sum_{n=2}^{\infty} \frac{f(n)}{E_1^{(0)} - E_n^{(0)}} a_0^2 = -2\mathcal{E}^2 a_0^3 \sum_{n=2}^{\infty} \frac{n^2 f(n)}{n^2 - 1} \qquad (5.153)$$

The summation in this equation is calculated to be $\cong 1.125$, therefore the second order energy shift becomes

$$E_{100}^2 \cong -2.25\mathcal{E}^2 a_0^3 \qquad (5.154)$$

This is known as the **quadratic Stark effect**. For any hydrogenlike atom with nuclear charge Z, $a_0 \to a_0/Z$.

The electric dipole moment **d** is obtained from the second order energy shift

$$d = -\frac{\partial E_{100}^{(2)}}{\partial \mathcal{E}} \cong 4.5 a_0^3 \mathcal{E} \qquad (5.155)$$

This is proportional to the electric field strength, that is, the dipole moment is induced. The **polarizability** is calculated from

$$P = \frac{d}{\mathcal{E}} \cong 4.5 a_0^3 \qquad (5.156)$$

Let us now calculate the first order energy shift of the excited states of the hydrogenlike atoms. Let us take $n = 2$; for the unperturbed system there are four $n = 2$ states that have the same energy, these are ϕ_{200}, ϕ_{211}, ϕ_{210}, ϕ_{21-1}. We will apply the degenerate perturbation theory for this case. We want to solve an equation like

$$\sum_i \alpha_i < \phi_n^{(j)}|H_1|\phi_n^{(i)} > = E_n^{(1)}\alpha_j \; ; \quad j = 1 - 4 \qquad (5.157)$$

This gives a 4×4 matrix equation. We have still $H_1 = e\mathcal{E}z$. Since some of the matrix elements vanish,

$$< \phi_{21\mp1}|z|\phi_{21\mp1} > = 0 \quad or \quad < 21 \mp 1|z|21 \mp 1 > = 0 \qquad (5.158)$$

the 4×4 matrix equation reduces to a 2×2 matrix equation. The resultant equation reads

$$e\mathcal{E} \begin{pmatrix} <200|z|200> & <200|z|210> \\ <210|z|200> & <210|z|210> \end{pmatrix} \begin{pmatrix} \alpha_1 \\ \alpha_2 \end{pmatrix} = E_2^{(1)} \begin{pmatrix} \alpha_1 \\ \alpha_2 \end{pmatrix}$$

(5.159)

The matrix elements are

$$<200|z|200> = <210|z|210> = 0 \qquad (5.160)$$

$$<210|z|200> = <200|z|210> = -3a_0$$

Therefore the matrix equation, Eq.(5.159), becomes

$$\begin{pmatrix} E_2^{(1)} & 3e\mathcal{E}a_0 \\ 3e\mathcal{E}a_0 & E_2^{(1)} \end{pmatrix} \begin{pmatrix} \alpha_1 \\ \alpha_2 \end{pmatrix} = 0 \qquad (5.161)$$

The eigenvalues of this are

$$E_2^{(1)} = \mp 3e\mathcal{E}a_0 \qquad (5.162)$$

This is known as the **linear Stark effect**. The corresponding eigenfunctions are $(\phi_{200} \pm \phi_{210})/\sqrt{2}$. Thus the linear Stark effect for $n = 2$ states of hydrogenlike atoms yields a splitting of degenerate levels as follows.

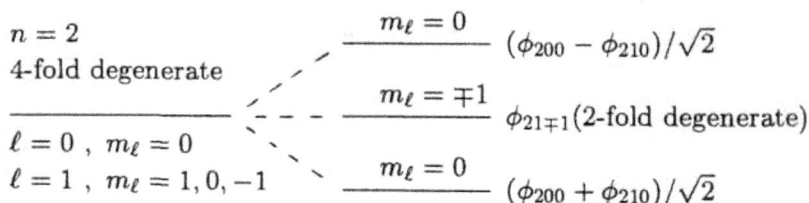

Fig. 5.21. Stark effect of hydrogen for $n = 2$.

5.19 Special hydrogenic systems

The key results we have obtained for hydrogenic systems are:
The energy eigenvalues:

$$E_n = -\frac{Z^2 e^2}{2n^2 a_\mu} \quad , \quad a_\mu = \frac{m}{\mu} a_0 \quad ; \quad E_n = -\frac{Z^2 e^2 \mu}{2n^2 m a_0} \quad (5.163)$$

a_μ is the modified Bohr radius.
The frequencies of the transitions:

$$\nu_{nn'} = Z^2 R(M) \left(\frac{1}{n'^2} - \frac{1}{n^2} \right) \quad , \quad R(M) = \frac{\mu}{m} R_\infty \quad (5.164)$$

In particular, the ionization potential $I_p = |E_{n=1}|$ is

$$I_p = \frac{Z^2 e^2}{2 a_\mu} = \frac{Z^2 e^2 \mu}{2 m a_0} = Z^2 \left(\frac{e^2 \mu}{2 m a_0} \right) \quad (5.165)$$

The **extension** a of the wave function describing the relative motion of the system in the ground state (most probable distance):

$$a = \frac{a_\mu}{Z} = \frac{\hbar^2}{Z e^3 \mu} = \frac{m a_0}{Z \mu} = \frac{1}{Z} \left(\frac{m a_0}{\mu} \right) \quad (5.166)$$

The hydrogenic systems we have considered so far correspond to an atomic nucleus of mass M and charge Ze, and an electron of mass m and charge $-e$ interacting by means of the coulomb potential. The **normal** hydrogen atom, (p, e^-), containing a proton and an electron is the prototype of these hydrogenic systems. The hydrogenic ions $He^+(Z = 2)$, $Li^{++}(Z = 3)$, $B^{3+}(Z = 4)$, etc., are also examples of such systems.

The value of a for these ions reduced with respect to that of the hydrogen atom by a factor of Z and their ionization potential is increased by a factor of Z^2.

The neutral isotopes of atomic hydrogen (H), deuterium (D) and tritium (T), also provide examples of hydrogenic systems; a deuteron (containing one proton and one neutron) in the case of deuterium or a triton (containing one proton and two neutrons) in the case of tritium.

Since $M_d \cong 2M_p$ and $M_t \cong 3M_p$, the reduced mass μ is slightly different for H, D, and T; the relative differences being of the order of 10^{-3}. Thus the quantities I_p and a are nearly identical for the three atoms, the small differences in the value of μ giving rise to isotopic shifts of the spectral lines (of the order of 10^{-3}).

Positronium, Muonium

Positronium (e^+, e^-) is a bound hydrogenic system made of a positron e^+ and an electron e^-, it is an isotope of hydrogen. It is unstable. Muonium (μ^+, e^-) is another isotope of hydrogen, in which the proton has been replaced by a positive muon μ^+, a particle which is very similar to the positron e^+, except that it has a mass $M_{\mu^+} \cong 207 m_e$ and that it is unstable, with a lifetime of about 2.2×10^{-6} s. Muonium is also unstable and its lifetime is the same as the muon.

Positronium and muonium contain only **leptons** (i.e., particles which are not affected by the strong interactions).

In positronium the electron and the positron may annihilate, their total energy including their rest mass energy being completely converted into electromagnetic radiation (photon): $e^+ + e^- \rightarrow \gamma(h\nu)$.

5.20   Exotic atoms

In all the hydrogenic atoms we have considered the negative particle is an electron. Other negative particles could form a bound system with a nucleus. These negative particles can be leptons such as the negative muon μ^- or **hadrons** (i.e., particles can have strong interactions). The unusual atoms formed in this way are sometimes called **exotic atoms**.

Muonic Atoms

Muonic atoms are formed when a negative muon μ^- is captured by the Coulomb attraction of a nucleus of charge Ze as the muon is slowing down in bulk matter.

The simplest muonic atom (p, μ^-) which contains a proton p and a muon μ^-. Since $M_{\mu^-} \cong 207m_e$, the reduced mass of the muon with respect to the proton is $\mu(p, \mu^-) \cong 186m_e$. As a result, the radius a of the muonic atom (p, μ^-) is 186 times smaller than that of the hydrogen atom, the ionization potential I_p of (p, μ^-) being 186 times larger than the corresponding quantity for the hydrogen atom. $a \approx a_H/186$, $I_p \approx 186 I_H$.

The frequencies of the spectral lines corresponding to transitions between the energy levels of (p, μ^-) may be obtained from those of the H–atom by multiplying by a factor of 186. For transitions between the lowest levels of (p, μ^-) the spectral lines are lying in the X–ray region. Low lying transitions are in X–ray region.

Assume that the negative muon μ^- is captured by the coulomb field of a nucleus of charge Ze. For this bound system $a \cong a_0/207Z$, the ionization potential I_p larger than that of hydrogen by a factor of $207Z^2$.

For the case of muonic lead $(Z = 82)$, we would have $I_p \cong 19 \ MeV$, and $a \cong 3 \times 10^{-15} \ m = 3 \ Fermi$. In fact this value of a is smaller than the radius r of the lead nucleus, which is $r \cong 6.7 \ Fermi$, so that the expressions obtained for hydrogen cannot be used any more. They have been derived on the assumption that the two particles of the hydrogenic system interact by means of the coulomb potential for all values of the relative distance r, i.e., both particles are considered to be point–like. This assumption is an excellent one for **usual** (electronic) atoms or ions such as H, D, T, He^+, Li^{++}, etc., where the finite extension of the nucleus gives rise to very small effect, the **volume effect**.

The volume effect causes tiny shifts of the energies associated with low–lying s–states. However, for muonic atoms with large values of Z (such as muonic lead) the volume effect may lead to important shifts of the low–lying levels (in particular of the $1s$ and $2s$ states).

The spectral lines corresponding to transitions between the lowest energy levels lie at the limit of the X–ray and γ–ray regions, and in the $1s$ state the muon spends a significant fraction of its time within the nucleus. The energy spectrum of muonic atoms is terefore sensitive to the internal structure of the nucleus constitutes.

Table 5.5: The reduced mass μ, radius of the first orbit a, and ionization potential I_p of some hydrogenic systems (in *a.u.*).

System	μ	a	I_p
(pe^-)	$\cong 1$	$\cong 1$	$\cong 0.5$
(e^+e^-)	0.5	2	0.25
(μ^+e^-)	$\cong 1$	$\cong 1$	$\cong 0.5$
$(p\mu^-)$	$\cong 186$	$\cong 5.4 \times 10^{-3}$	$\cong 93$
$(p\pi^-)$	$\cong 238$	$\cong 4.2 \times 10^{-3}$	$\cong 119$
(pK^-)	$\cong 633$	$\cong 1.6 \times 10^{-3}$	$\cong 317$
$(p\bar{p})$	$\cong 918$	$\cong 1.1 \times 10^{-3}$	$\cong 459$
$(p\Sigma^-)$	$\cong 1029$	$\cong 9.7 \times 10^{-4}$	$\cong 515$

Hadronic Atoms

In contrast with the leptons (such as e^-, e^+, μ^-, μ^+) which participate only in the EM and weak interactions, the hadrons participate in the strong (nuclear–type) interactions in addition to the EM and weak interactions.

There are two kinds of hadrons, the **baryons** (p, n, antiproton $\bar{p}$, antineutron $\bar{n}$, the hyperons Σ, Z, ...) which have half–integer spin $(1/2, 3/2, ...)$ and are therefore fermions, and the **mesons** (such as π–mesons, the K–mesons, etc.) which have integer spin $(0, 1, ...)$ and hence are bosons.

Among the hadrons, those having a negative charge can form with a nucleus N a **hydrogenic–type** system which is referred to as a **hadronic atom**. In particular the system (N, π^-) is called a **pionic atom**, (N, K^-) is known as a **kaonic atom** while $(N, \bar{p})$ is called an **antiprotonic atom**. Hydrogenic–type systems containing a nucleus and a negative hyperon, e.g. (N, Σ^-), are known as **hyperonic atoms**. All these hadronic atoms are unstable, but their lifetime is long enough so that some of their spectral lines have actually been observed.

Since hadrons inteact strongly with nuclei, the theory of hydrogenic system, which only takes into account the coulomb interaction, can not be directly applied to hadronic atoms. However, because strong interactions have a short range, exited states of hadronic atoms and in particular those with $\ell \neq 0$ for which the wave function is very small in the vicinity of the origin can essentially be studied. The energies of these exited states are thus given by $E_n \cong -I_p/n^2$ and their radii by $a_n \cong n^2 a$.

Rydberg Atoms

A highly excited atom (or ion) has an electron with a large principle quantum number n. The electron (or the atom) is said to be in a **high Rydberg state** and the highly excited atom is also referred to simply as a **Rydberg atom**. Some of the quantities of the H–atom with $n = 1$ and $n = 100$ are compared in Table 5.6.

It is clear from this table that higly excited H–atoms with $n = 100$ exhibit some remarkable properties. For example, their size is enormous on the atomic scale. The electron in a high Rydberg state is very weakly bound. The selective excitation of atoms in highly excited states requires experimental techniques with extremely high resolution. A highly excited H–atom left to itself has a relatively long lifetime. However, even thermal collisions can transfer enough energy to the atom to ionise it, although it is possible that a neutral system passing through the Rydberg atom will leave it undisturbed.

For a large enough n, a Rydberg atom of any kind may be considered as an **ionic core** plus a single highly excited electron. If this electron has enough angular momentum, so that it does not significantly penetrate the core, it will essentially move in a Coulomb field corresponding to an effective charge $Z_{eff} = 1$ (in a.u.). Such Rydberg atoms are therefore very similar to highly excited (Rydberg) hydrogen atom.

Table 5.6: Comparison of some characteristics of the H-atom for different values of the principal quantum number n.

Quantity	$n = 1$	arbitrary n	$n = 100$
Radius a_n of Bohr orbit (in m)	$a_0 \cong$ 5.3×10^{-11}	$= n^2 a_0$	5.3×10^{-7}
Geometric cross-section πa_n^2 (in m^2)	$\pi a_0^2 \cong$ 8.8×10^{-21}	$= n^4 \pi a_0^2$	8.8×10^{-13}
Binding energy $|E_n|$ (in eV)	$I_p^H \cong$ 13.6	$= I_p^H / n^2$	1.36×10^{-3}
Energy separation ΔE between adjacent levels (in eV)	large $n \;\rightarrow$	$\cong 2 I_p^H / n^3$	2.7×10^{-5}
RMS velocity of electron v_n (in m/s)	$v_0 = c\alpha \cong$ 2.2×10^6	$= v_0 / n$	2.2×10^4
Period T_n (in $second$)	$T_0 \cong$ 1.5×10^{-16}	$n^3 T_0$	1.5×10^{-10}
Electrical field $\mathcal{E}_n$ (in V/m)	$\mathcal{E}_0 \cong$ 5.1×10^{11}	$\mathcal{E}_0 / n^4$	5.1×10^3
Magnetic field B_n (in $Tesla$)	$B_0 \cong$ 2.4×10^5	B_0 / n^3	2.4×10^{-1}

5.21 Worked examples

Example - 5.1 :

Calculate the probability of finding a H–atom $1s$ electron outside the classically forbidden region.

Solution :

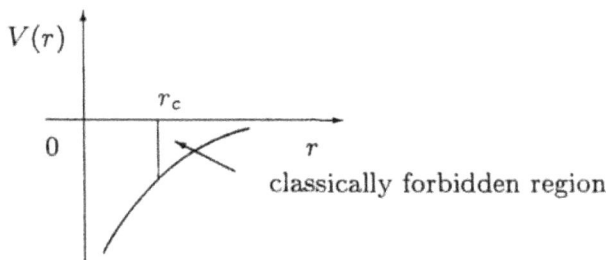

$$\psi_{1s} = \left(\frac{Z^3}{\pi a_0^3}\right)^{1/2} e^{-Zr/a_0} \quad \longrightarrow \quad \psi_{1s} = \frac{1}{\sqrt{\pi}} e^{-r} \quad (in \ a.u.)$$

$$V(r) = -\frac{Ze^2}{r_c} = E_{1s} = -\frac{Z^2 e^2}{2n^2 a_0} \quad \longrightarrow \quad -\frac{1}{r_c} = -\frac{1}{2} \quad \longrightarrow \quad r_c = 2 \ (in \ a.u.)$$

$$P_{1s}(r) = 4\pi r^2 \psi_{1s}^2(r)$$

$$P = \int_0^{r_c} P_{1s}(r) dr = \int_0^2 4\pi r^2 \frac{1}{\pi} e^{-2r} dr = 4 \int_0^2 r^2 e^{-2r} dr$$

$$= 4\left[\frac{2!}{2^3} - e^{-4} \sum_{k=0}^2 \frac{2!}{k!} \frac{2^k}{2^{2-k+1}}\right] = 4\left[\frac{1}{4} - 2e^{-4}\left(\frac{1}{2^3} + \frac{2}{2^2} + \frac{2^2}{2!2}\right)\right]$$

$$= 1 - 8e^{-4}\left(\frac{1}{8} + \frac{1}{2} + 1\right) = 1 - 13e^{-4} \cong 1 - 0.238 \cong 0.762$$

Here we used the following integrals:

$$\int_0^\infty x^n e^{-\alpha x} dx = \frac{n!}{\alpha^{n+1}} \ , \quad \int_0^a x^n e^{-\alpha x} dx = \frac{n!}{\alpha^{n+1}} - e^{-\alpha a} \sum_{k=0}^n \frac{n!}{k!} \frac{a^k}{\alpha^{n-k+1}}$$

Example - 5.2 :

Calculation of $3s$ wavefunction (ψ_{3s}):

Solution :

$$\psi_n(r) = N_n \left(\sum_{i=0}^{n-1} b_i r^i \right) e^{-ar}$$

$$\psi_3 = N_3(b_0 + b_1 r + b_2 r^2)e^{-ar} \equiv R_3(r)$$

$$\frac{dR_3(r)}{dr} = N_3(-ab_0 + b_1 - ab_1 r + 2b_2 r - ab_2 r^2)e^{-ar}$$

$$\frac{d^2 R_3(r)}{dr^2} = N_3(a^2 b_0 - 2ab_1 + a^2 b_1 r + 2b_2 - 4ab_2 r + a^2 b_2 r^2)e^{-ar}$$

Substitute in the radial Schrödinger equation:

$$-\frac{\hbar^2}{2m} \left(\frac{d^2 R(r)}{dr^2} + \frac{2}{r} \frac{dR(r)}{dr} \right) - \frac{Ze^2}{r} R(r) = ER(r)$$

$$-\frac{\hbar^2}{2m}(a^2 b_0 - 4ab_1 + a^2 b_1 r + 6b_2 - 6ab_2 r + a^2 b_2 r^2 - 2a\frac{b_0}{r} + \frac{2b_1}{r})$$

$$-Ze^2(\frac{b_0}{r} + b_1 + b_2 r) - E(b_0 + b_1 r + b_2 r^2) = 0$$

Separating with respect to the powers of r one gets the following equations:

coefficient of r^0: $-\frac{\hbar^2}{2m}(a^2 b_0 - 4ab_1 + 6b_2) - Ze^2 b_1 - Eb_0 = 0$

coefficient of r^1: $-\frac{\hbar^2}{2m}(a^2 b_1 - 6ab_2) - Ze^2 b_2 - Eb_1 = 0$

coefficient of r^{-1}: $-\frac{\hbar^2}{2m}(-2ab_0 + 2b_1) - Ze^2 b_0 = 0$

coefficient of r^2: $-\frac{\hbar^2}{2m}(a^2 b_2) - Eb_2 = 0$

Solving for the coefficients one gets:

$$a = \frac{Z}{3a_0} \quad , \quad b_0 = 27 \quad , \quad b_1 = -\frac{18Z}{a_0} \quad , \quad b_2 = \frac{2Z^2}{a_0^2}$$

$$R_3(r) = N_3 \left(27 - \frac{18Z}{a_0}r + 2(\frac{Z}{a_0})^2 r^2 \right) e^{-Zr/3a_0}$$

Considering the normalization condition: $\int_v R_3^2(r) dv = 1$

one gets the normalization constant: $N_3 = \frac{1}{81}\left(\frac{Z^3}{3\pi a_0^3}\right)^{1/2}$

Example - 5.3 :

Choosing the function $xyR(r)$ determine the wavefunction for $3d$ state.

Solution :

$$xyR(r) = (r\sin\theta\cos\phi)(r\sin\theta\sin\phi)R(r) = \psi(r,\theta,\phi)$$

Substituting this function in the Schrödinger equation

$$-\frac{\hbar^2}{2m}\left[\frac{1}{r^2}\frac{\partial}{\partial r}\left(r^2\frac{\partial}{\partial r}\right) + \frac{1}{r^2\sin\theta}\frac{\partial}{\partial\theta}\left(\sin\theta\frac{\partial}{\partial\theta}\right) + \frac{1}{r^2\sin^2\theta}\frac{\partial^2}{\partial\phi^2}\right]\psi - \frac{Ze^2}{r}\psi = E\psi$$

One gets the radial Schrödinger equation:

$$-\frac{\hbar^2}{2m}\left[\frac{d^2R(r)}{dr^2} + \frac{6}{r}\frac{dR(r)}{dr}\right] - \frac{Ze^2}{r}R(r) = ER(r)$$

Choose $R(r) = e^{-ar}$ and substitute in the radial equation, then one gets

$$a = \frac{Z}{3a_0} \text{ and } E = -\frac{1}{9}\frac{Z^2e^2}{2a_0}.$$

Normalization condition gives, $\psi(r,\theta,\phi) = NxyR(r)$

$$\int_v |\psi|^2 dv = 1 \quad\longrightarrow\quad N^2\int_v x^2y^2R^2(r)dv = 1$$

$$N = \frac{1}{81}\left(\frac{2}{\pi}\right)^{1/2}\left(\frac{Z}{a_0}\right)^{7/2} = \frac{1}{27}\sqrt{2/3}\left(\frac{Z^7}{3\pi a_0^7}\right)^{1/2}$$

$$\psi_{3d} = \frac{1}{81}\left(\frac{2}{\pi}\right)^{1/2}\left(\frac{Z}{a_0}\right)^{7/2}r^2\sin^2\theta\sin\phi\cos\phi e^{-Zr/3a_0}.$$

Example - 5.4 :

Consider a H–atom interacting with an external electric field

$E_z = E_z^0 \cos[2\pi(z/\lambda - \nu t)]$. Assume that only the two states $\psi_1 = |100>$ and $\psi_2 = |210>$ are coupled by the radiation. At $t = 0$ the atom is in the state ψ_1. Calculate the probability of finding the atom in the state ψ_2 at time t, and the intensity of the transition.

Solution :

$$P_2(t) \cong \frac{e^2(E_z^0)^2}{\hbar^2} |<\psi_1|x|\psi_1>|^2 t^2$$

$$<\psi_1|x|\psi_1> =$$

$$\int_0^\infty r^2 dr \int_0^\pi \sin\theta d\theta \int_0 2\pi d\phi N_2 \frac{Z}{a_0} e^{-Zr/2a_0} \cos\theta (r\sin\theta\cos\phi) N_1 e^{-Zr/a_0}$$

$$= N_1 N_2 \frac{Z}{a_0} \int_0^\infty r^4 e^{-3Zr/2a_0} dr \int_0^\pi \sin^2\theta\cos\theta d\theta \int_0^{2\pi} \cos\phi d\phi$$

$$= \frac{Z^3}{\pi a_0^3 \sqrt{2}} \frac{Z}{a_0} \left(4!(\frac{2a_0}{3Z})^5\right) \left(\frac{\sin^3\theta}{3}|_0^\pi\right) (\sin\phi|_0^{2\pi}) = 0$$

Therefore, $P_2(t) = 0$. Since $I \propto \frac{dP(t)}{dt} \quad \longrightarrow \quad I = 0$.

Example - 5.5 :

Assume that a H-atom is in an external uniform magnetic field along z-direction with field strength $1\ T$. Calculate the frequencies of the normal Zeeman triplet. Consider $2p \rightarrow 1s$ transition without spin-orbit interaction.

Solution :

$$\nu_+ = \frac{E_{2p} - E_{1s}}{h} + \frac{\mu_B}{h} B \qquad \Delta m = +1$$

$$\nu_+ = \frac{E_{2p} - E_{1s}}{h} \qquad\qquad \Delta m = 0$$

$$\nu_- = \frac{E_{2p} - E_{1s}}{h} - \frac{\mu_B}{h} B \qquad \Delta m = -1$$

$$\frac{E_{2p} - E_{1s}}{h} = \frac{1}{h}\left(-\frac{1}{4}\frac{Z^2 e^2}{2a_0} + \frac{Z^2 e^2}{2a_0}\right) = \frac{1}{h}\left(\frac{Z^2 e^2}{2a_0}\right)(1 - \frac{1}{4}) = \frac{3Z^2 e^2}{8a_0 h}$$

for H-atom $Z = 1$ $\longrightarrow$ $\frac{3e^2}{8a_0 h}$

in SI units:

$$\frac{3e^2}{8a_0 h} \longrightarrow \frac{3e^2}{8a_0 h (4\pi\varepsilon_0)} = \frac{3\times(1.6\times10^{-19})^2\times9\times10^9}{8\times0.53\times10^{-10}\times6.6\times10^{-34}} \, s^{-1} 1 \cong 2.47 \times 10^{15} \, s^{-1}$$

$$\frac{\mu_B}{h} B = \frac{9.27\times10^{-24}\times1}{6.6\times10^{-34}} \cong 1.4 \times 10^{10} \, s^{-1}$$

$$\nu_+ = (2.47 \times 10^{15} + 1.4 \times 10^{10}) \, s^{-1}$$

$$\nu_0 = 2.47 \times 10^{15} \, s^{-1}$$

$$\nu_- = (2.47 \times 10^{15} - 1.4 \times 10^{10}) \, s^{-1}$$

Example - 5.6 :

Estimate the separation in wave numbers for (a) the $2 \to 1$ transitions in hydrogen and deuterium (^2H), (b) the $2 \to 1$ transitions in hydrogen and tritium (^3H).

Solution :

a) For hydrogen : $\tilde{\nu}_{2\to1} = R_\infty \frac{\mu_H}{m_e} \frac{3}{4}$, $\mu = \frac{m_e m_p}{m_e + m_p}$

For deuterium : $\tilde{\nu}_{2\to1} = R_\infty \frac{\mu_d}{m_e} \frac{3}{4}$, $\mu = \frac{m_e m_d}{m_e + m_d}$

assume $m_p \approx m_n \approx 1838 m_e$, $m_d \cong 2m_p$

$$\lambda_H - \lambda_D = \left(\frac{1}{\tilde{\nu}_H} - \frac{1}{\tilde{\nu}_D}\right) = \frac{4}{3}\frac{m_e}{R_\infty}\left(\frac{m_p + m_e}{m_p m_e} - \frac{2m_p + m_e}{2m_p m_e}\right)$$

$$= \frac{4}{3}\frac{1}{R_\infty}\left(\frac{m_p + m_e}{m_p} - \frac{2m_p + m_e}{2m_p}\right) = \frac{4}{3}\frac{1}{R_\infty}\left(\frac{m_e}{2m_p}\right) = 3.30 \times 10^{-11} \, m = 0.330 \, \mathring{A}$$

b) Similarly

$$\lambda_H - \lambda_T = \frac{4}{3}\frac{1}{R_\infty}\left(\frac{m_p + m_e}{m_p} - \frac{3m_p + m_e}{3m_p}\right)$$

$$= \tfrac{4}{3}\tfrac{1}{R_\infty}\left(\tfrac{2m_e}{3m_p}\right) = 4.37 \times 10^{-11} \; m = 0.437 \; \text{Å}.$$

Example - 5.7 :

Find the possible term symbols for the $1s$, $2p$, $3d$, and $4f$ electrons.

Solution :

$1s:$ $(2S+1) = 2 \times \tfrac{1}{2} + 1 = 2$; $L = 0 \; (S)$; $J = 0 + \tfrac{1}{2} = \tfrac{1}{2}$

$$1(^2S_{\frac{1}{2}})$$

$2p:$ $(2S+1) = 2$; $L = 1 \; (P)$; $J = 1 + \tfrac{1}{2} , \; 1 - \tfrac{1}{2} \to \tfrac{3}{2} , \tfrac{1}{2}$

$$2(^2P_{\frac{3}{2}}) \; , \; 2(^2P_{\frac{1}{2}})$$

$3d:$ $(2S+1) = 2$; $L = 2 \; (D)$; $J = 2 + \tfrac{1}{2} , \; 2 - \tfrac{1}{2} \to \tfrac{5}{2} , \tfrac{3}{2}$

$$3(^2D_{\frac{5}{2}}) \; , \; 3(^2D_{\frac{3}{2}})$$

$4f:$ $(2S+1) = 2$; $L = 3 \; (F)$; $J = 3 + \tfrac{1}{2} , \; 3 - \tfrac{1}{2} \to \tfrac{7}{2} , \tfrac{5}{2}$

$$4(^2F_{\frac{7}{2}}) \; , \; 4(^2F_{\frac{5}{2}})$$

Example - 5.8 :

Consider a H–atom in an electric field so that the Hamiltonian becomes:
$H = H_0 + eE_z r \cos\theta$. Here the nucleus is placed at the origin and the
field E_z is in the z–direction. The Hamiltonian for the unperturbed
H–atom is H_0 which is assumed to be much larger than the perturbation. Use linear variation theory, i.e., $\psi = \sum_i c_i \phi_i$, with the functions
$\phi_1 = \phi_{1s} = |1s >$, $\phi_2 = \phi_{1p_z} = |1p_z >$, to determine the lowest energy
state for this atom and to estimate the polarizability.

Solution :

We are given the trial wavefunction $\psi = c_1\phi_1 + c_2\phi_2$. Minimization of the energy w.r. to c_1 and c_2 leads to the secular equation

$$\begin{vmatrix} H_{11} - ES_{11} & H_{12} - ES_{12} \\ H_{21} - ES_{21} & H_{22} - ES_{22} \end{vmatrix} = 0$$

$$H_{11} = < 1s|H_0|1s > +eE_z < 1s|z|1s >= E_{1s}$$

$$H_{22} = < 2p_z|H_0|2p_z > +eE_z < 2p_z|z|2p_z >= E_{1s}/4$$

$$H_{12} = eE_z < 1s|z|2p_z >$$

$$= \frac{eE_z a_0}{4\pi\sqrt{2}} \int_0^\infty e^{-3\sigma/2}\sigma^4 d\sigma \int_0^{2\pi} d\phi \int_0^\pi \cos^2\theta \sin\theta d\theta$$

$$= \frac{eE_z a_0}{\sqrt{2}} \frac{4!}{(3/2)^5} \frac{1}{3} = \frac{eE_z a_0}{\sqrt{2}} \frac{2^8}{3^5} = A$$

$$S_{11} = S_{22} = 1 \quad, \quad S_{12} = 0 \quad ; \quad \begin{vmatrix} E_{1s} - E & A \\ A & E_{1s}/4 - E \end{vmatrix} = 0$$

or $E^2 - (\frac{5}{4}E_{1s})E + (E_{1s}^2/4 - A^2) = 0$

The roots are $E = \frac{5}{8}E_{1s} \pm \frac{1}{2}\sqrt{(\frac{5}{4}E_{1s})^2 - 4(E_{1s}^2/4 - A^2)}$

or $E = \frac{5}{8}E_{1s} \pm \frac{1}{2}\sqrt{\frac{9}{16}E_{1s}^2(1 + \frac{16}{9E_{1s}^2}A^2)}$

Assuming that $A^2/E_{1s}^2 \ll 1$, we find $E = \frac{5}{8}E_{1s} \pm \frac{3}{8}E_{1s}(1 + \frac{32}{9}A^2/E_{1s}^2 + ...)$

The lowest energy is

$$E = E_{1s} + \frac{4}{3}\frac{A^2}{E_{1s}} = E_{1s} + \frac{4}{3}a_0^2(\frac{2^{16}}{3^{10}})\frac{e^2 E_z^2}{2E_{1s}} = E_{1s} - (2.96)\frac{E_z^2}{2}a_0^3(4\pi\varepsilon_0)$$

here we used $E_{1s} = -\frac{e^2}{(4\pi\varepsilon_0)2a_0}$.

Since the energy of polarization is given by $E_{polar} = -\frac{\alpha}{2}E_z^2$, from the above expression one gets $\alpha = 2.96a_0^3(4\pi\varepsilon_0)$. The exact value of α for the H–atom is $\frac{9}{2}a_0^3(4\pi\varepsilon_0)$.

Example - 5.9 :

The spin–orbit interaction for the electron in a hydrogen–like atom is given by: $H_{so} = \xi(r)\mathbf{L} \cdot \mathbf{S}$ where $\xi(r) = \frac{\hbar^2}{2m^2c^2}(\frac{1}{r}\frac{\partial V}{\partial r})$. Derive an equation for the allowed energies in such atoms in terms of the quantum numbers ℓ, s, j.

Solution :

The complete Hamiltonian is $H = H_0 + H_{so}$, where

$H_0|n\ell> = E_0(n\ell)|n\ell>$.

$\mathbf{J} = \mathbf{L} + \mathbf{S}$ and $J^2 = L^2 + S^2 + 2\mathbf{L} \cdot \mathbf{S}$ $\rightarrow$ $\mathbf{L} \cdot \mathbf{S} = \frac{1}{2}(J^2 - L^2 - S^2)$

In the representation $|n\ell> \rightarrow |\ell s j m_j>$,

$<\ell s j m_j|\mathbf{L} \cdot \mathbf{S}|\ell s j m_j> = \frac{1}{2}[j(j+1) - \ell(\ell+1) - s(s+1)]$

The complete energy expression is:

$E(n, \ell, s, j) = <n\ell s j m_j|H_0 + H_{so}|n\ell s j m_j>$

$= E_0(n\ell) + <n\ell|\xi(r)|n\ell><\ell s j m_j|\mathbf{L} \cdot \mathbf{S}|\ell s j m_j>$

Defining $\zeta(n\ell) = <n\ell|\xi(r)|n\ell>$, $E(n\ell s j) = E_0(n\ell) + E_{so}$

$E_{so} = \frac{\zeta(n\ell)}{2}[j(j+1) - \ell(\ell+1) - s(s+1)]$

Since $s = 1/2$, $E(n\ell s j) = E_0(n\ell) + \zeta(n\ell)\frac{\ell}{2}$, $j = \ell + \frac{1}{2}$

$= E_0(n\ell) - \zeta(n\ell)\frac{(\ell+1)}{2}$, $j = \ell - \frac{1}{2}$

Evaluation of $\zeta(n\ell)$:

$\zeta(n\ell) = \frac{\hbar^2}{2m^2c^2} <n\ell|\frac{1}{r}\frac{\partial V}{\partial r}|n\ell> = \frac{\hbar^2}{2m^2c^2} <n\ell|\frac{1}{r}\frac{\partial}{\partial r}(\frac{-Ze^2}{r})|n\ell>$

$$= \frac{Zh^2 e^2}{2m^2 c^2} < n\ell|\frac{1}{r^3}|n\ell > = \frac{Zh^2 e^2}{2m^2 c^2}\left[\frac{Z^3}{a_0^3 n^3 \ell(\ell+\frac{1}{2})(\ell+1)}\right]$$

$$= \frac{(Ze)^4 m}{2h^2}\left(\frac{e^2}{ch}\right)^2 \frac{1}{n^3 \ell(\ell+\frac{1}{2})(\ell+1)}$$

Example - 5.10 :

Find the relativistic correction for the energy levels of hydrogen–like atoms using first order perturbation theory. Assume that $v \ll c$ and obtain the correction to the order of $(v/c)^2$.

Solution :

We start with $E^2 = p^2 c^2 + m_0^2 c^4$, which is the relativistic energy of the electron, with rest mass m_0. The corresponding Hamiltonian may be written as $H = \sqrt{p^2 c^2 + m_0^2 c^4} - e\phi$, where ϕ is the electrostatic potential. Since $(v/c)^2 \ll 1$,

$$H = m_0 c^2 \sqrt{\frac{p^2}{m_0^2 c^2} + 1} - e\phi = m_0 c^2\left(1 + \frac{p^2}{2m_0^2 c^2} - \frac{p^4}{8m_0^4 c^4} + ...\right) - e\phi$$

$$= m_0 c^2 + \frac{p^2}{2m_0} - \frac{p^4}{8m_0^3 c^3} + ... - e\phi$$

Choosing $m_0 c^2$ as the reference energy we have:

$$W = (H - m_0 c^2) = (\frac{p^2}{2m_0} - e\phi) - \frac{p^4}{8m_0^3 c^3} = H_0 + H_1$$

H_0 gives the non–relativistic eigenvalue for the hydrogen–like atom, and H_1 is a small perturbation. Thus $H_0|n\ell m >= E_n^{(0)}|n\ell m >$

$$E^{(1)} =< n\ell m|H_1|n\ell m >= -\frac{1}{2m_0 c^2} < n\ell m|\frac{p^4}{4m_0^2}|n\ell m >$$

From $\frac{p^2}{2m_0} = E_n^{(0)} + e\phi \quad \rightarrow \quad p^2 = 2m_0(E_n^{(0)} + e\phi)$

and $E^{(1)} = -\frac{1}{2m_0 c^2} < n\ell m|(E_n^{(0)} + e\phi)^2|n\ell m >$ (in Rydberg units)

$$= -\frac{1}{2m_0 c^2}\left(\frac{2\pi^2 m_0 e^4}{h^2}\right) < n\ell m|(-\frac{Z^2}{n^2} + \frac{2Z}{r})^2|n\ell m >$$

$$= -\frac{\alpha^2}{4}\left[\frac{Z^4}{n^4} - \frac{4Z^3}{n^2}<\frac{1}{r}> +4Z^2<\frac{1}{r^2}>\right] \quad , \quad \alpha = \frac{e^2}{\hbar c}$$

Substituting $<\frac{1}{r}>_{n\ell m}$ and $<\frac{1}{r^2}>_{n\ell m}$ we obtain

$$E^{(1)} = -\frac{\alpha^2}{4}\left[\frac{Z^4}{n^4} - \frac{4Z^4}{n^4} + \frac{4Z^4}{n^3(\ell+\frac{1}{2})}\right] = -\frac{\alpha^2 Z^4}{n^3}\left[\frac{1}{(\ell+\frac{1}{2})} - \frac{3}{4n}\right]$$

Example - 5.11 :

Calculation of the $C - G$ coefficients for the coupling of two spins $1/2$:

Solution :

System   possible states

| 1 | $|1/2>, |-1/2>$ |
| 2 | $|-1/2>, |1/2>$ |

$1+2$: $|1/2>|1/2>$, $|1/2>|-1/2>$, $|-1/2>|1/2>$, $|-1/2>|-1/2>$

or : $|10>$, $|11>$, $|1-1>$, $|00>$

The relation between them is given by the $C - G$ coefficients.

It is clear that $|11>= |1/2>|1/2>$

Applying the step operator J_- find $|10>$,

$J_-|11>= \sqrt{2}|10>$

$J_-|11>= (j_{1_-} + j_{2_-})|1/2>|1/2>= |-1/2>|1/2> +|1/2>|-1/2>$

Therefore, $|10>= \frac{1}{\sqrt{2}}(|-1/2>|1/2> +|1/2>|-1/2>)$

$J_-|10>= \sqrt{2}|1-1>$

$(j_{1-} + j_{2-}) \frac{1}{\sqrt{2}} (| - 1/2 > |1/2 > + |1/2 > | - 1/2 >)$

$= \frac{1}{\sqrt{2}} (| - 1/2 > | - 1/2 > + | - 1/2 > | - 1/2 >) = \sqrt{2} | - 1/2 > | - 1/2 >$

Therefore, $|1 - 1 > = | - 1/2 > | - 1/2 >$

Considering the orthogonality relation among the states $|10 >$, $|11 >$, $|1 - 1 >$, and $|00 >$ one can easily find that

$|00 > = \frac{1}{\sqrt{2}} (|1/2 > | - 1/2 > - | - 1/2 > |1/2 >)$

Another approach:

$s_1 = 1/2$, $m_{s_1} = +1/2$, $-1/2$; $s_2 = 1/2$, $m_{s_2} = +1/2$, $-1/2$

Possible states : (I)

$|s_1 m_{s_1}, s_2 m_{s_2} >$ $\rightarrow$ $|m_{s_1}, m_{s_2} >$

$|1/2\ 1/2, 1/2\ 1/2 >$ $\rightarrow$ $|+, + >$
$|1/2\ 1/2, 1/2\ -1/2 >$ $\rightarrow$ $|+, - >$
$|1/2\ -1/2, 1/2\ 1/2 >$ $\rightarrow$ $|-, + >$
$|1/2\ -1/2, 1/2\ -1/2 >$ $\rightarrow$ $|-, - >$

For total spin: $\mathbf{S} = \mathbf{s_1} + \mathbf{s_2}$, $S = 1, 0$, $M_S = 1, 0, -1; 0$

Possible states : (II)

$|SM_S > :$ $|11 >$, $|10 >$, $|1 - 1 >$, $|00 >$

Relations between (I) and (II) :

$|11 > = |+, + >$

$|10 > = \frac{1}{\sqrt{2}} (|+, - > + |-, + >)$

$|1 - 1 > = |-, - >$

$|00> = \frac{1}{\sqrt{2}}(|+,-> -|-,+>)$

so we have the matrix equation:

$$\begin{pmatrix} |11> \\ |10> \\ |1-1> \\ |00> \end{pmatrix} = \begin{pmatrix} 1 & 0 & 0 & 0 \\ 0 & \frac{1}{\sqrt{2}} & \frac{1}{\sqrt{2}} & 0 \\ 0 & 0 & 0 & 1 \\ 0 & \frac{1}{\sqrt{2}} & -\frac{1}{\sqrt{2}} & 0 \end{pmatrix} \begin{pmatrix} |+,+> \\ |+,-> \\ |-,+> \\ |-,-> \end{pmatrix}$$

Example - 5.12 :

How many terms are obtained when the ground state level of a hydrogen atom splits due to interaction of the magnetic moments of the electron and proton (hyperfine splitting)? What is the order of the energy difference ΔE between the subshells of the hyperfine structure?

Solution :

Hyperfine splitting in hydrogen leads to the formation of two sublevels from one level. The order of magnitude is

$$\Delta E \sim \frac{\mu_e \mu_p}{r^3} \sim 10^{-18} \; erg$$

Here the magnetic moments of the electron and proton are:

$$\mu_e = \frac{eh}{2mc} = 9 \times 10^{-21} \; erg/oersted$$

$$\mu_p \cong \frac{3eh}{2Mc} \sim 1.4 \times 10^{-23} \; erg/oersted$$

and $r \sim 5 \times 10^{-9}$ cm is the radius of the first Bohr orbit.

This estimate leads to too low a value of ΔE since the electron comes closer to the nucleus in actual fact than according to Bohr's theory. When $r \sim 2 \times 10^{-9}$ cm the estimate agrees well with an accurate calculation.

Chapter 6

Many–Electron Atoms

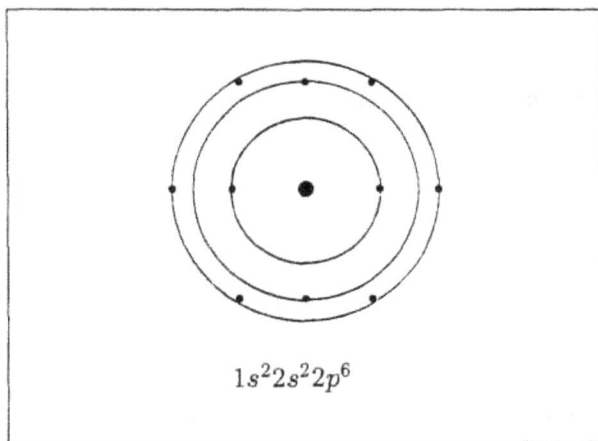

$$1s^2 2s^2 2p^6$$

6.1   The Pauli exclusion principle

For the two–electron helium atom the ground state wave function may be approximated by

$$\psi_{He}(1,2) = 1s(1)1s(2) \tag{6.1}$$

we might suppose that we could handle in the same manner atoms with more than two electrons. For example, for the ground state of the three–electron lithium atom, let us put all the electrons into the lowest energy $1s$ orbital $1s(1)1s(2)1s(3)$. This function is not acceptable even as a first approximation. The elimination of such functions is required by a basic quantum–mechanical principle, called the **Pauli exclusion principle.** W. Pauli first introduced it in 1925, and he postulated that **no two electrons can have the same set of quantum numbers.**

For atomic electrons, the quantum numbers n, ℓ, m_ℓ, and m_s; the fifth quantum number, s, which gives the magnitude of the spin, is $1/2$ for all electrons.

We do not violate the exclusion principle by writing the helium ground state as Eq.(6.1), even though n, ℓ, m_ℓ have the same value for the two electrons, m_s can be $+1/2$ for one electron and $-1/2$ for the other.

The symbols α and β are used for the spin functions corresponding to $m_s = +1/2$ and $-1/2$, respectively.

$$S_z|\chi_{s,m_s}> = m_s\hbar|\chi_{s,m_s}> \tag{6.2}$$

$$S_z|\alpha> = +(1/2)|\alpha> \quad , \quad S_z|\beta> = -(1/2)|\beta>$$

The functions α and β are orthonormal and they do not depend upon continuous variables.

$$<\alpha|\alpha> = <\beta|\beta> = 1 \quad , \quad <\alpha|\beta> = 0 \tag{6.3}$$

For neglegible interaction between the orbital and spin magnetic moments, the space and spin parts of the wave function are independent, the total wave function can be written

$$\psi_{He}(1,2) = 1s\alpha(1)1s\beta(2) \tag{6.4}$$

and the Pauli principle is obeyed.

Although the wave function given in Eq.(6.4) obeys the Pauli principle, it is not completely satisfactory because it distinguishes between the electrons. However, all electrons have identical properties and are indistinguishable.

A function that is identical to the function given in Eq.(6.4) is

$$\psi_{He}(2,1) = 1s\alpha(2)1s\beta(1) \tag{6.5}$$

the electron labels are interchanged.

If we add the two functions in Eqs.(6.4) and (6.5), we obtain

$$\psi_{He}(1,2) + \psi_{He}(2,1) = 1s\alpha(1)1s\beta(2) + 1s\alpha(2)1s\beta(1) \tag{6.6}$$

The square of this function (probability distribution) is

$$[\psi_{He}(1,2) + \psi_{He}(2,1)]^2 = [1s\alpha(1)]^2[1s\beta(2)]^2 \tag{6.7}$$

$$+[1s\alpha(2)]^2[1s\beta(1)]^2 + 2[1s\alpha(1)1s\beta(2)1s\alpha(2)1s\beta(1)]$$

Interchanging the labels in Eq.(6.7) gives the same distribution, Eq.(6.7). Thus, it is impossible to say which electron is which.

Eq.(6.6) is not the only way of combining the functions $\psi_{He}(1,2)$ and $\psi_{He}(2,1)$ to obtain a probability distribution that does not distinguish between the electrons. Another possibility is to take the difference and write

$$\psi_{He}(1,2) - \psi_{He}(2,1) = 1s\alpha(1)1s\beta(2) - 1s\alpha(2)1s\beta(1) \tag{6.8}$$

which gives the probability distribution

$$[\psi_{He}(1,2) - \psi_{He}(2,1)]^2 = [1s\alpha(1)]^2[1s\beta(2)]^2 \tag{6.9}$$

$$+[1s\alpha(2)]^2[1s\beta(1)]^2 - 2[1s\alpha(1)1s\beta(2)1s\alpha(2)1s\beta(1)]$$

Here again, interchanging the labels of the two electrons in Eq.(6.9) gives the same distribution.

The functions themselves, Eqs.(6.6) and (6.8), do not both have the property that they are unaltered by interchanging electrons. The function, Eq.(6.6), does not change by interchanging electrons

$$\psi_{He}(1,2) + \psi_{He}(2,1) = \psi_{He}(2,1) + \psi_{He}(1,2) \qquad (6.10)$$

Therefore, we say that such a function is **symmetric** for electron interchange. By contrast, the function, Eq.(6.8), changes sign by interchanging electrons

$$\psi_{He}(1,2) - \psi_{He}(2,1) = -[\psi_{He}(2,1) - \psi_{He}(1,2)] \qquad (6.11)$$

The function in Eq.(6.8) is said to be **antisymmetric** to electron interchange. Thus, there appear to be two different ground state helium wave functions, which give different probability distributions. However, all particles with half–integer spin (fermions) are described by antisymmetric wave functions, and all particles with zero or integer spin (bosons) are described by symmetric wave functions.

Therefore, the properly antisymmetrized helium–atom ground state is represented by

$$\psi(1,2) = \frac{1}{\sqrt{2}}[1s\alpha(1)1s\beta(2) - 1s\alpha(2)1s\beta(1)] \qquad (6.12)$$

$$= \frac{1}{\sqrt{2}}1s(1)1s(2)[\alpha(1)\beta(2) - \alpha(2)\beta(1)]$$

The factorization of spatial and spin parts is possible only for two–electron systems.

Another way of writing $\psi(1,2)$ is in the equivalent determinantal form,

$$\psi(1,2) = \frac{1}{\sqrt{2}} \begin{vmatrix} 1s\alpha(1) & 1s\beta(1) \\ 1s\alpha(2) & 1s\beta(2) \end{vmatrix} \qquad (6.13)$$

Eq.(6.13) is called a **Slater determinant**, after J.C. Slater, who introduced this convenient form for electronic wave functions.

The rows of the determinant are labelled by the electron indices $(1, 2)$ and the columns by the forms of the one–electron wave functions

$(1s\alpha, 1s\beta)$, which are commonly called **spin orbitals**. A Slater determinantal wave function automatically satisfies the Pauli principle. An alternative statement of the Pauli exclusion principle is the following: **A wave function for a system of two or more electrons must be antisymmetric with respect to interchange of the labels of any two electrons.**

Let us consider again the ground state of the lithium atom. If we put all three electrons in a $1s$ orbital, that is, $n_1 = n_2 = n_3 = 1$, $\ell_1 = \ell_2 = \ell_3 = 0$, and $m_{\ell_1} = m_{\ell_2} = m_{\ell_3} = 0$, three different values of m_s would be required to satisfy the Pauli principle. Since m_s may take only the two values $\pm 1/2$, the Pauli principle excludes the function $1s(1)1s(2)1s(3)$. Thus, to obtain a satisfactory wave function, at least one electron must be in an orbital with higher energy.

For hydrogen, the next orbitals have $n = 2$ and all of the $n = 2$ orbitals $(2s, 2p_0, 2p_{+1}, 2p_{-1})$ have the same energy; they are said to be degenerate. However, the shielding effects due to the presence of more than one electron in an atom cause the energy of the $2s$ orbital to be somewhat lower than that of the $2p$ orbital. Thus, in the ground state of lithium, there are two electrons in the $1s$ orbital (with spins **paired**; i.e., $m_{s_1} = +1/2, m_{s_2} = -1/2$) and one electron in the $2s$ orbital. The antisymmetrized approximate wave function can be written in two different forms,

$$\psi_{Li}(1,2,3) = \frac{1}{\sqrt{6}} \begin{vmatrix} 1s\alpha(1) & 1s\beta(1) & 2s\alpha(1) \\ 1s\alpha(2) & 1s\beta(2) & 2s\alpha(2) \\ 1s\alpha(3) & 1s\beta(3) & 2s\alpha(3) \end{vmatrix} \qquad (6.14)$$

or

$$\psi_{Li}(1,2,3) = \frac{1}{\sqrt{6}} \begin{vmatrix} 1s\alpha(1) & 1s\beta(1) & 2s\beta(1) \\ 1s\alpha(2) & 1s\beta(2) & 2s\beta(2) \\ 1s\alpha(3) & 1s\beta(3) & 2s\beta(3) \end{vmatrix}$$

Both of these functions have the same energy in the isolated atom, since in the absence of a magnetic field a $2s$ orbital with α spin is degenerate with one of β spin.

The assignment of the electrons in an atom to the available orbitals is called the **electron configuration**. We often specify the configuration simply in terms of the occupied orbitals with or without explicit designation of the spin. Thus, for the ground state of helium, we can write $1s\alpha1s\beta$ or $1s^2$ and for the ground state of lithium $1s\alpha1s\beta2s\alpha$ or $1s^22s$ and so on.

6.2 Many–electron determinantal wave function

As a short hand for an N–electron wave function, it is often expressed as a product of N spin orbitals

$$\psi_N = \psi(1, 2, ..., N) = \prod_{i=1}^{N} S_i(i) \tag{6.15}$$

This form of N–electron wave function is called as the **Hartree product**, it is not antisymmetric with respect to the exchange of a pair of electron coordinates. The antisymmetrized form is the Slater determinantal form

$$det(\psi_N) = \frac{1}{\sqrt{N!}} \begin{vmatrix} S_1(1) & S_2(1) & \dots & S_N(1) \\ S_1(2) & S_2(2) & \dots & S_N(2) \\ \vdots & \vdots & & \vdots \\ S_1(N) & S_2(N) & \dots & S_N(N) \end{vmatrix} \tag{6.16}$$

each spinorbital is assumed to be separately normalized. The product of the diagonal elements is just the Hartree product, the remaining $(N!-1)$ spinorbital products represent all possible permutations of the electron coordinates among the spinorbitals. The determinantal wave function can be written in the symbolic form

$$det(\psi_N) = \mathcal{A}\psi_N \tag{6.17}$$

where $\mathcal{A}$ is the antisymmetrization operator. For an N–electron wave function in the orbital approximation $\mathcal{A}$ has the general form

$$\mathcal{A} = \frac{1}{\sqrt{N!}} \sum_{\mathcal{P}} (-1)^{\mathcal{P}} \mathcal{P} \tag{6.18}$$

This operator converts a product of N doubly indexed elements into a determinantal form. The operator carries out $N!$ different permutations of the electron coordinates to form the components of an antisymmetric wave function

$$\sum_{\mathcal{P}} (-1)^{\mathcal{P}} \mathcal{P} = 1 - \sum_{ij} \mathcal{P}_{ij} + \sum_{ijk} \mathcal{P}_{ijk} - \dots \qquad (6.19)$$

where $\mathcal{P}_{ij}$, $\mathcal{P}_{ijk}$ are the two–particle and three–particle permutation operators, respectively. Permutations are carried out only upon the subscripts of the spin orbitals.

For example, for a three–electron wave function, if $\psi_3 = \prod_{i=1}^{3} S_i(i)$

$$\mathcal{A}\psi_3 = \frac{1}{\sqrt{3!}} [S_1(1) S_2(2) S_3(3) - S_2(1) S_1(2) S_3(3) \qquad (6.20)$$

$$-S_3(1) S_2(2) S_1(3) - S_1(1) S_3(2) S_2(3) + S_2(1) S_3(2) S_1(3) + S_3(1) S_1(2) S_2(3)]$$

Some properties of A :

The permutation operators are unitary,

$$\mathcal{P}^+ = \mathcal{P}^{-1} \qquad (6.21)$$

The antisymmetrization operator is self–adjoint,

$$\mathcal{A}^+ = \mathcal{A} \qquad (6.22)$$

$$\mathcal{A}^2 = \sqrt{N!}\mathcal{A} \qquad (6.23)$$

If $\mathcal{H}$ is a spin–free N–electron hamiltonian

$$[\mathcal{H}, \mathcal{A}] = 0 \qquad (6.24)$$

that is $\mathcal{H}$ commutes with $\mathcal{A}$.

6.3 Two–electron atoms

The principles of the quantum theory for many–electron atoms are the same as those already described for the hydrogen atom. However, the electrostatic interactions of the electrons with one another, as well as with the nucleus, makes accurate calculations of the wave functions and energy levels for many–electron atoms much more difficult. Nevertheless, it is possible to develop qualitative concepts for understanding atomic structure and atomic spectra. This is done by introducing suitable approximations for the wave function and energies of the atomic electrons.

Ground state of the Helium atom:

Helium atom can be considered as the simplest many–electron atom. It consists of a nucleus with charge $+2e$ and two electrons.

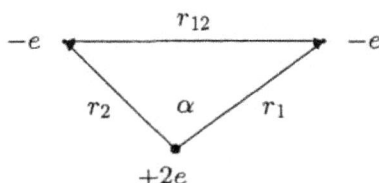

Fig. 6.1. Relative positions of the particles in He atom.

According to Coulomb's law, the potential energy is

$$V(r_1, r_2, r_{12}) = -\frac{2e^2}{r_1} - \frac{2e^2}{r_2} + \frac{e^2}{r_{12}} \tag{6.25}$$

where the first and second terms correspond to the attraction between the nucleus and electrons 1 and 2, respectively, and the third term arises from the mutual repulsion of the two electrons. To write the Schrödinger equation for He atom we assume that the nucleus is stationary because of its large mass relative to that of the electrons, and we write kinetic energy terms for the electrons only. The resulting Schrödinger equation is

$$\left(-\frac{\hbar^2}{2m}\nabla_1^2 - \frac{\hbar^2}{2m}\nabla_2^2 \right)\psi + V(r_1, r_2, r_{12})\psi = E\psi \tag{6.26}$$

where the Laplacian operator is given by

$$\nabla_i^2 = \frac{\partial^2}{\partial x_i^2} + \frac{\partial^2}{\partial y_i^2} + \frac{\partial^2}{\partial z_i^2} \quad , \quad i = 1, 2 \tag{6.27}$$

E represents the total energy of the electrons in the atom and ψ is the electronic wave function

$$\psi = \psi(x_1, y_1, z_1; x_2, y_2, z_2) \tag{6.28}$$

We can write the Eq.(6.26) in a more compact form

$$H\psi = E\psi \tag{6.29}$$

where the operator H (Hamiltonian operator) has the form

$$H = -\frac{\hbar^2}{2m}\nabla_1^2 - \frac{\hbar^2}{2m}\nabla_2^2 + V(r_1, r_2, r_{12}) \tag{6.30}$$

The problem is now to find the helium atom energy E and the associated wave functions ψ which satisfy Eq.(6.26). Because the potential energy term depends on the radial distances r_1 and r_2 it is again convenient to use spherical polar coordinates. We express the wave function as

$$\psi = \psi(r_1, \theta_1, \phi_1; r_2, \theta_2, \phi_2) = \psi(1, 2) \tag{6.31}$$

The operator ∇_i^2 can be written as

$$\nabla_i^2 = \frac{1}{r_i^2}\frac{\partial}{\partial r_i}\left(r_i^2\frac{\partial}{\partial r_i}\right) + \frac{1}{r_i^2\sin\theta}\frac{\partial}{\partial\theta_i}\left(\sin\theta_i\frac{\partial}{\partial\theta_i}\right) + \frac{1}{r_i^2\sin^2\theta}\frac{\partial^2}{\partial\phi_i^2} \tag{6.32}$$

In contrast to the H–atom Schrödinger equation, the He–atom equation, Eq.(6.26), has no exact solution. The difficulty comes from the electron–electron repulsion term e^2/r_{12} in the potential energy. It is necessary, therefore, to use approximation methods.

6.4   Independent–electron approximation

In this approximation the electron–electron repulsion term (e^2/r_{12}) is neglected which couples the motion of the two electrons. The resulting equation has the form

$$\left(-\frac{\hbar^2}{2m}\nabla_1^2 - \frac{2e^2}{r_1}\right)\psi + \left(-\frac{\hbar^2}{2m}\nabla_2^2 - \frac{2e^2}{r_2}\right)\psi = E\psi \tag{6.33}$$

Each of the terms in parenthesis is the Hamiltonian operator for a hydrogen–like atom with nuclear charge $+2e$. In this case each electron moves independently, therefore the wave function can be written as the product of two functions

$$\psi(1.2) = \psi_1(1)\psi_2(2) \tag{6.34}$$

The factors $\psi_1(1)$ and $\psi_2(2)$ satisfy the one–electron atom equations

$$\left(-\frac{\hbar^2}{2m}\nabla_1^2 - \frac{2e^2}{r_1}\right)\psi_1 = E_1\psi_1 \quad , \quad \left(-\frac{\hbar^2}{2m}\nabla_2^2 - \frac{2e^2}{r_2}\right)\psi_2 = E_2\psi_2 \quad (6.35)$$

Substitution of Eq.(6.34) into Eq.(6.33) with the use of Eq.(6.35) gives

$$(E_1 + E_2)\psi_1(1)\psi_2(2) = E\psi_1(1)\psi_2(2) \tag{6.36}$$

Total energy E is the sum of two one–electron energies

$$E = E_1 + E_2 \tag{6.37}$$

The ground state energy is then obtained as

$$E_0 = E_1 + E_2 = E_{1s} + E_{1s} = 2E_{1s}(Z=2) = 2\left(-\frac{Z^2e^2}{2a_0}\right) \tag{6.38}$$

$$= -8\left(\frac{e^2}{2a_0}\right) = -8 \times 13.6 \ eV = -108.8 \ eV$$

$$= -8 \times 0.5 \ a.u. = -4.0 \ a.u.$$

The true He–ground state energy is $-79.0 \ eV$.

The functions ψ_1 and ψ_2 corresponding to this energy are the hydrogen–like $1s$ functions ($Z=2$) (both electrons are in the $1s$ orbital)

$$\psi_1(1) = 1s(1) = \left(\frac{2^3}{\pi a_0^3}\right)^{1/2} e^{-2r_1/a_0} \tag{6.39}$$

$$\psi_2(2) = 1s(2) = \left(\frac{2^3}{\pi a_0^3}\right)^{1/2} e^{-2r_2/a_0}$$

For the first exited state of He the same approximation yields the energy

$$E_1 = E_{1s} + E_{2s} \tag{6.40}$$

and the wave function

$$\psi(1,2) = \psi_{1s}(1)\psi_{2s}(2) \equiv 1s(1)2s(2) \tag{6.41}$$

and so on. In general, for singly excited states $(1sn\ell)$, energy and wave function may be expressed as

$$E = E_{1s} + E_{n\ell} \quad , \quad \psi(1,2) = \psi_{1s}(1)\psi_{n\ell}(2) = 1s(1)n\ell(2) \tag{6.42}$$

6.5   Average–shielding approximation

In this approximation the effect of $e - e$ repulsion term is included in a simple way. Consider the behaviour of one of the electrons, say, electron 2, in the presence of the other. Electron 2 sees not only the nuclear charge $+2e$, but also the field due to electron 1.

In the ground state of the atom, electron 1 is in a $1s$ orbital, and it has a spherically symmetric probability distribution.

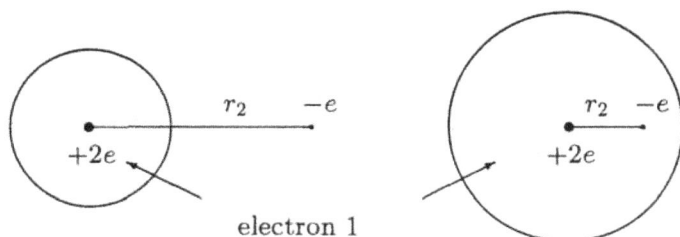

electron 1

Fig. 6.2. Average shielding models in He atom.

If electron 2 is far from the nucleus, it will be attracted by the nucleus and repelled by the negative cloud around the nucleus due to electron 1. The result is that electron 2 is attracted to the nucleus somewhat less than it would be in the one–electron He^+ ion; that is, the nuclear charge is **shielded** by the part of the electron cloud corresponding to electron 1. If r_2 is very large ($r_2 \rightarrow \infty$), the shielding result in an effective nuclear charge Z_e acting on electron 2 that is near unity.

On the other hand, if electron 2 is close to the nucleus, the shielding is much less complete. In the limit $r_2 \rightarrow 0$, the effective nuclear charge Z_e approaches the unshielded nuclear charge of 2. Thus, Z_e has the limits $1 \leq Z_e \leq 2$. In general, for any two–electron atom with nuclear charge Z, $(Z - 1) \leq Z_e \leq Z$.

The shielded wave functions are

$$\psi_1(1) = \left(\frac{Z_e^3}{\pi a_0^3}\right)^{1/2} e^{-Z_e r_1/a_0} \quad , \quad \psi_2(2) = \left(\frac{Z_e^3}{\pi a_0^3}\right)^{1/2} e^{-Z_e r_2/a_0} \qquad (6.43)$$

the same value of Z_e appears in both functions.

For a hydrogen–like atom with nuclear charge $+Z_e e$, the energy for the ground state $(1s)$ would be

$$E_{1s} = -\frac{Z_e^2 e^2}{2a_0} \qquad (6.44)$$

Each electron in He–atom has an energy given by Eq.(6.44). The first ionization potential, IP_1, can be approximated by

$$IP_1 = -E_{1s} = \frac{Z_e^2 e^2}{2a_0} = 13.6 Z_e^2 \ eV \qquad (6.45)$$

The second IP is exact, since there is no shielding of the nucleus (one–electron system)

$$IP_2 = \frac{Z^2 e^2}{2a_0} = 4 \times 13.6 \ eV = 54.4 \ eV. \qquad (6.46)$$

The experimental measurements give $(IP_1)_{exp} = 24.6 \ eV$, $(IP_2)_{exp} = 54.4 \ eV$. If we choose Z_e such that $(IP_1)_{exp} = Z_e^2 e^2/(2a_0)$, we find that $Z_e = 1.34$, it lies between the required limits 1 and 2.

The radial distribution of $1s$ function for various Z values behave as shown in the following figure.

In the shielding approximation wave functions are improved by using experimental energy value to obtain the effective nuclear charge.

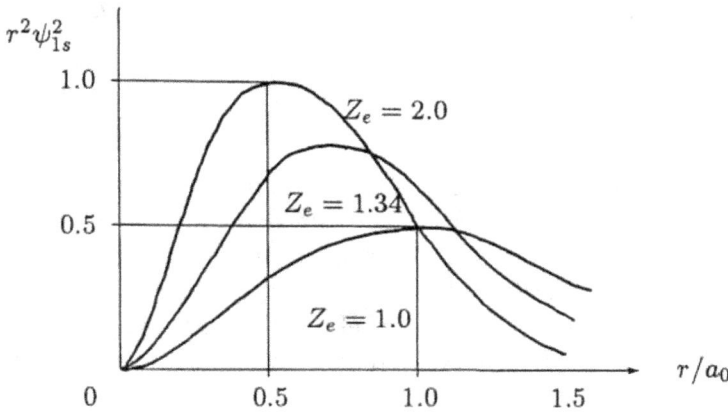

Fig. 6.3. Radial distribution of $1s$ function for various Z_e.

6.6 Perturbation approach

We assume that the helium ground state wave function is given by
Eq.(6.34), $\psi(1,2) = \psi_1(1)\psi_2(2)$, where

$$\psi_1(1) = 1s(1) = \left(\frac{2^3}{\pi a_0^3} \right)^{1/2} e^{-2r_1/a_0} \tag{6.47}$$

$$\psi_2(2) = 1s(2) = \left(\frac{2^3}{\pi a_0^3} \right)^{1/2} e^{-2r_2/a_0}$$

The average contribution of the potential energy term to the total energy
can then be calculated as

$$< V >= \int \psi^2(1,2)V(r_1, r_2, r_{12})dv_1 dv_2 \tag{6.48}$$

Substitution of Eqs.(6.34), (6.47), and (6.25) in (6.48) gives

$$< V >= \int \psi^2(1)\psi^2(2) \left(-\frac{2e^2}{r_1} - \frac{2e^2}{r_2} + \frac{e^2}{r_{12}} \right) dv_1 dv_2 \tag{6.49}$$

$$= \int \psi^2(1)\left(-\frac{2e^2}{r_1}\right) dv_1 + \int \psi^2(2)\left(-\frac{2e^2}{r_2}\right) dv_2$$

$$+ \int \psi^2(1)\psi^2(2)\left(\frac{e^2}{r_{12}}\right) dv_1 dv_2$$

The first term of Eq.(6.49) represents the attractive interaction between the charge distribution $-e\psi_1^2(1)$ and the point nucleus of charge $+2e$; the second term represents the corresponding interaction for electron 2 and the nucleus. The third term representing the interaction between the charge distribution $-e\psi_1^2(1)$ for electron 1 and $-e\psi_2^2(2)$ for electron 2, may be considered as a perturbation correction to the total energy using unshielded wave functions.

The nucleus–electron terms are the same as for the one–electron hydrogen–like atoms, $< V >_{n\ell m_\ell} = -Z^2 e^2/(n^2 a_0)$.

$$\int \psi_1^2(1)(-2e^2/r_1)dv_1 = \int \psi_2^2(2)(-2e^2/r_2)dv_2$$

Thus

$$< V >= 2\left(-\frac{4e^2}{a_0}\right) + \int \psi^2(1)\psi^2(2)\left(\frac{e^2}{r_{12}}\right) dv_1 dv_2 \qquad (6.50)$$

The $e - e$ term (after some algebra) can be calculated as

$$\int \psi^2(1)\psi^2(2)\left(\frac{e^2}{r_{12}}\right) dv_1 dv_2 = \frac{5}{2}\left(\frac{e^2}{2a_0}\right) \qquad (6.51)$$

Therefore, total ground state energy of helium atom becomes

$$E_{He} = E_1 + E_2 + E_{12} = -4\left(\frac{e^2}{2a_0}\right) - 4\left(\frac{e^2}{2a_0}\right) + \frac{5}{2}\left(\frac{e^2}{2a_0}\right) \qquad (6.52)$$

$$= -\frac{11}{2}\left(\frac{e^2}{2a_0}\right) = -\frac{11}{2} \times 13.6 \ eV = -74.8 \ eV$$

$$= -\frac{11}{2} \times 0.5 \ a.u. = -2.75 \ a.u.$$

$(E_{He})_{exp} = -79.0 \ eV$. In general,

$$E_n = -(Z^2/n^2)(e^2/2a_0) \quad , \quad E_{12} = (5/4)Z(e^2/2a_0)$$

$$E(two \ electron \ ground \ state) = \left(-2Z^2 + \frac{5}{4}Z\right)\left(\frac{e^2}{2a_0}\right) \qquad (6.53)$$

It is clear that introduction of the e^2/r_{12} term gives a considerable improved result.

The first ionization potential, IP_1, is

$$IP_1 = E_{He^+} - E_{He} = -54.4 - (-74.8) = 20.4 \ eV \qquad (6.54)$$

$(IP_1)_{exp} = 24.6 \ eV$.

6.7   The variation method

The variation method is based on a generalization of the minimization procedure by which the best Z_e was determined. For the ground state of an atom or molecule, the average value of the exact Hamiltonian operator for an approximate wave function is always greater than an exact energy. This important result is called the **variation principle**. The best form for an approximate wave function is the one which minimizes the average energy. This makes it possible to introduce any number of parameters into a wave function. These parameters are then varied simultaneously so as to find the lowest energy. Many of the more accurate calculations have been carried out in this way.

Consider the ground state of the He–atom. The exact non-relativistic hamiltonian is given by (in *a.u.*)

$$H = -\frac{1}{2}(\nabla_1^2 + \nabla_2^2) - Z(\frac{1}{r_1} + \frac{1}{r_2}) + \frac{1}{r_{12}} \qquad (6.55)$$

As a trial wave function we shall use a product of two hydrogen–like $1s$ orbitals with a modified nuclear charge, namely (in *a.u.*)

$$\psi = \left(\frac{\eta^3}{\pi}\right)^{1/2} e^{-\eta r_1} \left(\frac{\eta^3}{\pi}\right)^{1/2} e^{-\eta r_2} = \frac{\eta^3}{\pi}e^{-\eta(r_1+r_2)} \qquad (6.56)$$

where the modified nuclear charge η serves as a variational factor, which is also known as **effective nuclear charge** or **scale parameter**.

The expectation value of the Hamiltonian turns out to be

$$< H >=< \psi|H|\psi > \qquad (6.57)$$

$$=< \psi| - \frac{1}{2}(\nabla_1^2 + \nabla_2^2)|\psi > + < \psi| - Z(\frac{1}{r_1} + \frac{1}{r_2})|\psi > + < \psi|\frac{1}{r_{12}}|\psi >$$

$$= \eta^2 - 2Z\eta + \frac{5}{8}\eta \quad (in\ a.u.)$$

The variational requirement for a minimum in the energy is $\partial E/\partial \eta = 0$.

$$2\eta - 2Z + \frac{5}{8} = 0 \quad \rightarrow \quad \eta = Z - \frac{5}{16} \qquad (6.58)$$

For $Z = 2$, $\eta = 27/16 = 1.6875$. Therefore

$$E =< H >= -2.848\ a.u. = -77.5\ eV$$

When $\eta = Z = 2 \quad \rightarrow \quad E = -2.75\ a.u. = -74.8\ eV$ (unshielded wave function result).

Linear variation functions:

An atomic wave function may be expressed as a linear combination of basis functions (trial wave function)

$$\psi = \sum_{i=1}^{N} C_i\phi_i \qquad (6.59)$$

where C_i are the linear parameters. If ψ is not normalized

$$E = \frac{< \psi|H|\psi >}{< \psi|\psi >} = \frac{\sum_i \sum_j C_i C_j < \phi_i|H|\phi_j >}{\sum_i \sum_j C_i C_j < \phi_i|\phi_j >} \qquad (6.60)$$

$$= \frac{\sum_i \sum_j C_i C_j H_{ij}}{\sum_i \sum_j C_i C_j S_{ij}}$$

This is the approximate ground state energy. Variation principle requires that

$$\frac{\partial E}{\partial C_k} = 0 \quad , \quad k = 1, 2, ..., N \qquad (6.61)$$

The resulting equation (**Secular equation**) is obtained as

$$\sum_i C_i(H_{ik} - ES_{ik}) = 0 \quad , \quad k = 1, 2, ..., N \qquad (6.62)$$

N simultaneous homogeneous linear equations in the N independent variables $(C_1, C_2, ..., C_N)$. To have a non–trivial solution, determinant of coefficients (**Secular determinant**) must vanish.

$$|H_{ik} - ES_{ik}| = 0 \qquad (6.63)$$

The lowest root corresponds to the ground state energy, which is an upper limit to the exact energy E_0 ($E > E_0$).

Wave functions considered so far contain electron coordinates r_1 and r_2 only. Hylleraas introduced $e - e$ distance r_{12} into the wave function and obtained a better improvement in the ground state energy of helium.

Hylleraas variables:

$$r_1 + r_2 = s \quad , \quad r_2 - r_1 = t \quad , \quad r_{12} = u \qquad (6.64)$$

A trial wave function in terms of Hylleraas variables

$$\phi(s, t, u) = e^{-\alpha s}[1 + f(s, t, u)] \qquad (6.65)$$

where $f(s, t, u)$ is a power series in the variables s, t, and u. For example, the function investigated by Hylleraas

$$\phi = e^{-\alpha s}(1 + C_1 u + C_2 t^2 + C_3 s + C_4 s^2 + C_5 u^2) \qquad (6.66)$$

where α, and $\{C_i\}$, $i = 1 - 5$ are variational parameters, gives an energy of -2.90324 a.u. $(= -78.98\ eV)$, the experimental value (to the same number of significant figure) is -2.90372 a.u. $(= -79.0\ eV)$.

Comparison of energies for He ground state:

– Independent electron approximation:

$$E = 2E_{1s} = 2(-Z^2 e^2/2a_0) = -Z^2/2 \ a.u. = -4.0 \ a.u. = -108.8 \ eV$$

– Average shielding approximation:

Does not give energy, it is fitted to experimental value to determine Z_e in the wave functions. Fit to IP_1.

$$IP_1 = E_{He^+} - E_{He} = -E_{1s} = Z^2 e^2/2a_0$$

$$(IP_1)_{exp} = 24.6 \ eV = 13.6 Z_e^2 \quad \rightarrow \quad Z_e \cong 1.34$$

– Perturbation (with unshielded wave function):

$$E = (-2Z^2 + 5Z/4)(e^2/2a_0) = -Z^2 + 5Z/8 \ a.u.$$

$$E = -2.750 \ a.u. = -74.8 \ eV$$

– Variation (with shielded wave function):

(including 1 term) $\psi = (\eta^3/\pi)e^{-\eta(r_1+r_2)}$,

$$E = \eta^2 - 2Z\eta + 5\eta/8 \ a.u. \quad , \quad \eta = 1.6875 \ ,$$

$$E = -2.848 \ a.u. = -77.5 \ eV$$

– Variation (with Hylleraas wave function):

(including 6 terms) $\psi = e^{-\alpha s}(1 + f(s,t,u))$,

$$E = -2.90324 \ a.u. = -78.98 \ eV$$

– Experimental result:

$$E = -2.90372 \ a.u. = -79.0 \ eV.$$

The non–relativistic Hamiltonian operator for two–electron atoms is given (in *a.u.*) as

$$H = -\frac{1}{2}(\nabla_1^2 + \nabla_2^2) - Z(\frac{1}{r_1} + \frac{1}{r_2}) + \frac{1}{r_{12}} \tag{6.67}$$

The coordinates used in this Hamiltonian (r_1, r_2, r_{12}) are known as metric coordinates. The relation among them is

$$r_{12}^2 = r_1^2 + r_2^2 - 2r_1 r_2 \cos\theta \tag{6.68}$$

where θ is the angle between $\mathbf{r_1}$ and $\mathbf{r_2}$. A suitable wave function in metric coordinates could be $\psi = \psi(r_1, r_2, r_{12})$.

For S states (total orbital angular momentum is zero) the two–electron non–relativistic Hamiltonian operator is given in metric coordinates (in *a.u.*) as

$$H = -\frac{1}{2}\left[\frac{\partial^2}{\partial r_1^2} + \frac{2}{r_1}\frac{\partial}{\partial r_1} + \frac{\partial^2}{\partial r_2^2} + \frac{2}{r_2}\frac{\partial}{\partial r_2} + 2\frac{\partial^2}{\partial r_{12}^2} + \frac{4}{r_{12}}\frac{\partial}{\partial r_{12}}\right] \tag{6.69}$$

$$-\frac{1}{2}\left[\frac{r_1^2 + r_{12}^2 - r_2^2}{r_1 r_{12}}\frac{\partial^2}{\partial r_1 \partial r_{12}} + \frac{r_2^2 + r_{12}^2 - r_1^2}{r_2 r_{12}}\frac{\partial^2}{\partial r_2 \partial r_{12}}\right]$$

$$-Z\left(\frac{1}{r_1} + \frac{1}{r_2}\right) + \frac{1}{r_{12}}$$

The corresponding volume element is given by

$$dv = r_1^2 r_2^2 \sin\theta \, dr_1 dr_2 d\theta = r_1 r_2 r_{12} dr_1 dr_2 dr_{12} \tag{6.70}$$

If one uses a Hylleraas type wave function, $\psi = \psi(s, t, u)$, where $s = r_1 + r_2$, $t = r_2 - r_1$, and $u = r_{12}$, then one should express the Hamiltonian operator in terms of the Hylleraas variables (s, t, u). The two–electron non–relativistic Hamiltonian operator (in *a.u.*) and the corresponding volume element in Hylleraas coordinates for S states may be expressed as follows:

$$H = -\left(\frac{\partial^2}{\partial s^2} + \frac{\partial^2}{\partial t^2} + \frac{\partial^2}{\partial u^2}\right) - \frac{4s}{s^2 - t^2}\frac{\partial}{\partial s} + \frac{4t}{s^2 - t^2}\frac{\partial}{\partial t} - \frac{2}{u}\frac{\partial}{\partial u} \tag{6.71}$$

$$-2\frac{s}{u}\left(\frac{u^2-t^2}{s^2-t^2}\right)\frac{\partial^2}{\partial s \partial u} - 2\frac{t}{u}\left(\frac{u^2-s^2}{s^2-t^2}\right)\frac{\partial^2}{\partial t \partial u} - 4Z\frac{s}{s^2-t^2} + \frac{1}{u}$$

$$dv = \frac{1}{8}u(s^2-t^2)dsdtdu. \tag{6.72}$$

The range of s, t, u are:

$$0 \le t \le u \le s \le \infty \tag{6.73}$$

6.8   Excited states of Helium

If the (e^2/r_{12}) term in the Hamiltonian operator is ignored, the Schrödinger equation takes the form

$$(H_1 + H_2)\psi(1,2) = E\psi(1,2) \tag{6.74}$$

where

$$H_i = -\frac{\hbar^2}{2m}\nabla_i^2 - \frac{Ze^2}{r_i} \quad , \quad i = 1, 2 \tag{6.75}$$

and its solutions take the form

$$\psi(1,2) = \psi_1(1)\psi_2(2) \tag{6.76}$$

We would obtain the first excited state $1s2s$ by setting $\psi_1(1) = \psi_{1s}(1)$ and $\psi_2(2) = \psi_{2s}(2)$. The energy would be

$$E = E_{1s} + E_{2s} \tag{6.77}$$

The wave function $\psi_{1s}(1)\psi_{2s}(2)$ gives different distribution functions for the two electrons, it violates the principle of indistinguishability and is unacceptable. By analogy with the helium ground state we can form linear combinations of $\psi(1,2)$ and $\psi(2,1)$ such that the electrons are indistinguishable. The symmetric and antisymmetric linear combinations are

$$\psi_S = \psi(1,2) + \psi(2,1) = 1s(1)2s(2) + 1s(2)2s(1) \tag{6.78}$$

$$\psi_A = \psi(1,2) - \psi(2,1) = 1s(1)2s(2) - 1s(2)2s(1)$$

According to Pauli principle only ψ_A is a satisfactory description of the $He(1s2s)$ excited state. However, this form is not complete, because functions in Eq.(6.78) include only the spatial part of the wave function and the spin must be introduced to obtain a complete description of the state.

Considering the spin functions separately, we have the four possibilities:

$$\alpha(1)\alpha(2) , \; \beta(1)\beta(2) , \; \alpha(1)\beta(2) , \; \beta(1)\alpha(2)$$

or requiring indistinguishability of the electrons with respect to spin:

$$\alpha(1)\alpha(2) , \; \beta(1)\beta(2) , \; \alpha(1)\beta(2) + \beta(1)\alpha(2) , \; \alpha(1)\beta(2) + \beta(1)\alpha(2)$$

All four spin functions are possible for this case; the first three are symmetric with respect to electron interchange and the fourth is antisymmetric. To combine these with the spatial functions in such a way that the complete wave function is antisymmetric, we must multiply the symmetric spatial function by antisymmetric spin functions and vice versa; that is

$$\psi_1(1,2) = \frac{1}{\sqrt{2}}[1s(1)2s(2) + 2s(1)1s(2)]\frac{1}{\sqrt{2}}[\alpha(1)\beta(2) - \beta(1)\alpha(2)] \quad (6.79)$$

$$\psi_2(1,2) = \frac{1}{\sqrt{2}}[1s(1)2s(2) - 2s(1)1s(2)]\frac{1}{\sqrt{2}}[\alpha(1)\beta(2) + \beta(1)\alpha(2)]$$

$$\psi_3(1,2) = \frac{1}{\sqrt{2}}[1s(1)2s(2) - 2s(1)1s(2)]\alpha(1)\alpha(2)$$

$$\psi_4(1,2) = \frac{1}{\sqrt{2}}[1s(1)2s(2) - 2s(1)1s(2)]\beta(1)\beta(2)$$

It is also possible to write these functions in the determinantal form,

$$\psi_1(1,2) = \frac{1}{\sqrt{2}}\left\{ \frac{1}{\sqrt{2}} \begin{vmatrix} 1s\alpha(1) & 2s\beta(1) \\ 1s\alpha(2) & 2s\beta(2) \end{vmatrix} - \frac{1}{\sqrt{2}} \begin{vmatrix} 1s\beta(1) & 2s\alpha(1) \\ 1s\beta(2) & 2s\alpha(2) \end{vmatrix} \right\}$$

$$\psi_2(1,2) = \frac{1}{\sqrt{2}}\left\{ \frac{1}{\sqrt{2}} \begin{vmatrix} 1s\alpha(1) & 2s\beta(1) \\ 1s\alpha(2) & 2s\beta(2) \end{vmatrix} + \frac{1}{\sqrt{2}} \begin{vmatrix} 1s\beta(1) & 2s\alpha(1) \\ 1s\beta(2) & 2s\alpha(2) \end{vmatrix} \right\}$$

$$\psi_3(1,2) = \frac{1}{\sqrt{2}} \begin{vmatrix} 1s\alpha(1) & 2s\alpha(1) \\ 1s\alpha(2) & 2s\alpha(2) \end{vmatrix} \qquad (6.80)$$

$$\psi_4(1,2) = \frac{1}{\sqrt{2}} \begin{vmatrix} 1s\beta(1) & 2s\beta(1) \\ 1s\beta(2) & 2s\beta(2) \end{vmatrix}$$

To a good approximation the energy of each state depends only upon the form of the orbital part of the wave function associated with it; hence ψ_2, ψ_3, and ψ_4 are degenerate, and are said to form a **spin triplet** (multiplicity of three) while ψ_1 is nondegenerate and is said to be a **spin singlet** (multiplicity of one). We denote the orbital parts of the singlet and triplet wave functions by $^1\psi$ and $^3\psi$, respectively;

$$^1\psi(1,2) = \frac{1}{\sqrt{2}}[1s(1)2s(2) + 2s(1)1s(2)] = \frac{1}{\sqrt{2}}\psi_S \qquad (6.81)$$

$$^3\psi(1,2) = \frac{1}{\sqrt{2}}[1s(1)2s(2) - 2s(1)1s(2)] = \frac{1}{\sqrt{2}}\psi_A$$

If (e^2/r_{12}) term is omitted from the Hamiltonian, both $^1\psi$ and $^3\psi$ have energy $E_{1s} + E_{2s}$. When this term is included, the energies are different. The average value of (e^2/r_{12}) for $^1\psi$ is

$$< \frac{e^2}{r_{12}} >_S = \int \int {}^1\psi(1,2)(\frac{e^2}{r_{12}})^1\psi(1,2)dv_1dv_2 \qquad (6.82)$$

$$= \frac{e^2}{2} \int \int [1s(1)2s(2) + 2s(1)1s(2)]^2(\frac{1}{r_{12}})dv_1dv_2$$

$$= \frac{e^2}{2} \int \int [1s(1)]^2(\frac{1}{r_{12}})[2s(2)]^2 dv_1dv_2 + \frac{e^2}{2} \int \int [2s(1)]^2(\frac{1}{r_{12}})[1s(2)]^2 dv_1dv_2$$

$$+ e^2 \int \int [1s(1)2s(1)](\frac{1}{r_{12}})[1s(2)2s(2)]dv_1dv_2 = J + K$$

where

$$J = e^2 \int \int [1s(1)]^2(\frac{1}{r_{12}})[2s(2)]^2 dv_1dv_2 \qquad (6.83)$$

$$= e^2 \int \int [2s(1)]^2(\frac{1}{r_{12}})[1s(2)]^2 dv_1dv_2$$

$$K = e^2 \int \int [1s(1)2s(1)](\frac{1}{r_{12}})[1s(2)2s(2)]dv_1dv_2$$

For the function $^3\psi$ one can show similarly that

$$< \frac{e^2}{r_{12}} >_T = J - K \tag{6.84}$$

Thus the energies of the singlet and triplet states are

$$^1E = E_0 + J + K \quad , \quad ^3E = E_0 + J - K \tag{6.85}$$

where $E_0 = E_{1s} + E_{2s}$.

The integral J is called an **atomic Coulomb integral**, it represents the electrostatic coulomb interaction of the $1s$ and $2s$ charge clouds. The integral K is called an **atomic exchange integral**, it arises from an exchange of the two electrons between the $1s$ and $2s$ orbitals.

The degeneracy of $^1\psi$ and $^3\psi$ is split by the exchange interaction as shown in the figure

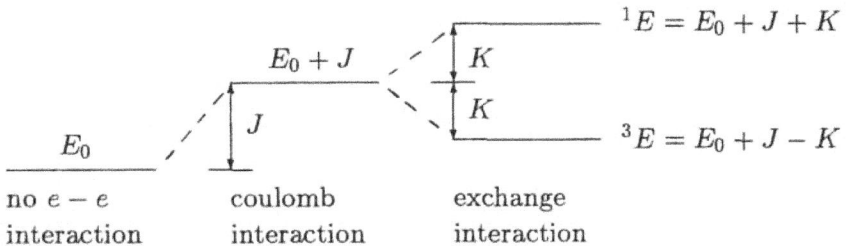

Fig. 6.4. Contribution of coulomb and exchange interactions.

The triplet energy is lower than the singlet energy by the amount $2K$. This ordering of the singlet and triplet levels arising from the assignment of the electrons to two different orbitals holds for all of the excited states of helium.

6.9   Para– and Ortho–Helium

In helium, the singlet states (called parahelium) and the triplet states (called orthohelium) may be treated as separate forms of the element.

A transition between states of different multiplicity requires a change of spin. Singlet to triplet $(S \rightarrow T)$ and triplet to singlet $(T \rightarrow S)$ transitions are very improbable and are said to be **forbidden**. Multiplicities greater than unite (i.e., spin degeneracies) indicate that an atom or molecule has a permanent magnetic dipole moment due to the unpaired electron spin. Substances containing such atoms or molecules are attracted by a magnetic field and are said to be **paramagnetic**.

If the multiplicity is unity (i.e., spin zero), no permanent magnetic moment exists. Substances without a permanent magnetic moment are repelled by magnetic fields and are said to be **diamagnetic**.

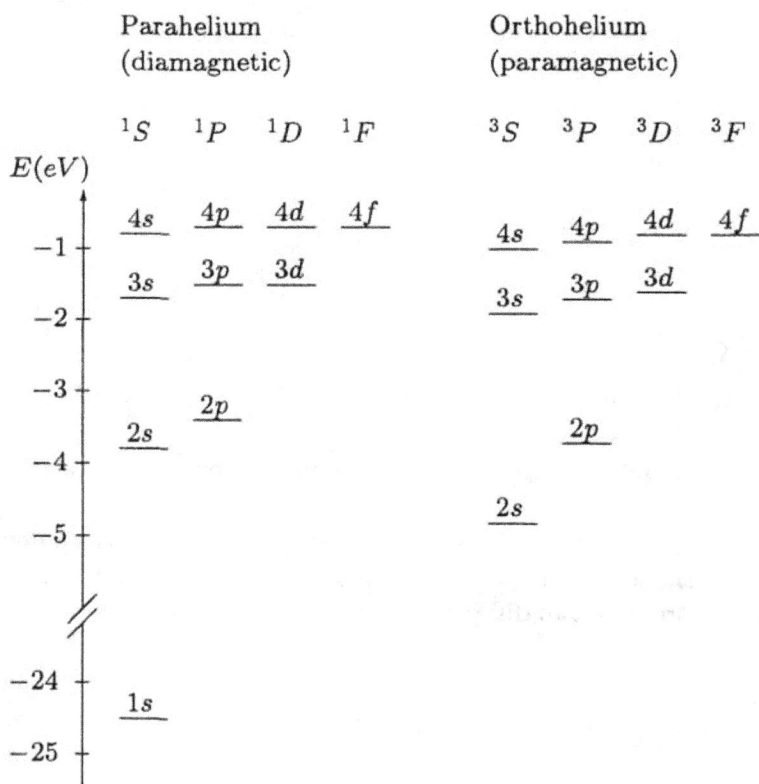

Fig. 6.5. Relative positions of helium energy levels.

6.10   Doubly excited Helium states

Each of the excited helium states in the previous diagram is a singly excited state; that is, one electron occupies the $1s$ ground state orbital and one occupies an excited orbital; $\{(1sn\ell);\ n > 1\}$. Doubly excited states in which both electrons occupy excited orbitals, $\{(n\ell n'\ell');\ n > 1, n' \geq n\}$, e.g., $He(2s^2)$, have energies greater than the first ionization limit. Thus, these doubly excited states of helium can undergo so called **autoionization** transitions forming a He^+ ion and a free electron.

The presence of such autoionizing states in the (He^+, e^-) continuum can have an observable effect on the absorption spectrum of the system.

Autoionization takes place as a result of double excitation. Two electrons are excited simultaneously. When the excitation energy of the two electrons exceeds the first *I.P.*, the atom may undergo a **radiationless** transition in which one of the excited electrons falls into a lower state, the de–excitation energy being used to eject the other excited electron from the atom.

This process occurs very quickly following double excitation. The lifetime of the excited atom against autoionization is generally very much shorter ($10^{-16} - 10^{-12}$ *second*) than its radiative lifetime ($10^{-9} - 10^{-7}$ *s*). If double excitation occurs for inner electrons then the autoionization process is called as **Auger effect**.

For neutral atoms $\longrightarrow$ autoionization
For atomic ions (negative) $\longrightarrow$ autodetachment
For molecules $\longrightarrow$ predissociation, preionization

All these processes are generally called as **resonances**

Importance and applications of autoionization:

– The time–independent Schrödinger equation does not give solutions for autoionizing states.
– Transitions between normal atomic levels and upper levels lying in the continuum are of astrophysical interest.

— Dielectronic recombination:

$$e^- + A^+ \longleftrightarrow A + h\nu \tag{6.86}$$

— Plasma physics.
— Surface science and chemistry (Auger spectroscopy).
— Isotope separation.

Primary processes for autoionization:

Positive autoionization by photon impact

$$A + h\nu \to A^{**} \quad , \quad A^{**} \to A^+ + e^- \tag{6.87}$$

Positive autoionization by electron impact

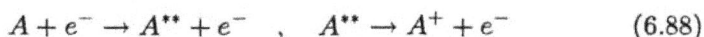

$$A + e^- \to A^{**} + e^- \quad , \quad A^{**} \to A^+ + e^- \tag{6.88}$$

Direct electron impact negative ionization, negative autoionization

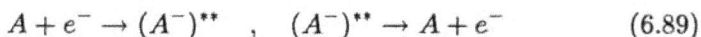

$$A + e^- \to (A^-)^{**} \quad , \quad (A^-)^{**} \to A + e^- \tag{6.89}$$

Electron impact positive ionization and excitation, autoionization of a positive excited ion (Auger effect)

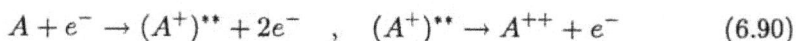

$$A + e^- \to (A^+)^{**} + 2e^- \quad , \quad (A^+)^{**} \to A^{++} + e^- \tag{6.90}$$

Methods to study autoionization:

Experimental:

— Cloud chamber (Auger effect).
— Continuous ultraviolet absorption spectroscopy.
— Electron scattering.

Theoretical:

— Fescbach projection operator formalism.
— Multiconfiguration energy bound method.

- Scattering close–coupling method.
- Multichannel configuration interaction method.
- Root–stabilization method.
- Complex–coordinate method.
- Truncated diagonalization method.
- Many–body perturbation theory.
- Time stability theory.
- Derivative method.

Resonance widths and line–shapes:

U. Fano (in 1961) defined the total wave function of an atomic system as

$$|\Psi> = a|\phi> + \int b(E)|\psi(E)> dE \qquad (6.91)$$

The first term on the right hand side describes the discrete region, while the second term describes the continuum of the system. According to Fano the line–width of a resonance state is given by

$$\Gamma = 2\pi| <\psi(E)|V_{12}|\phi> |^2 \qquad (6.92)$$

The potential V_{12} represents the transition operator and is called the coupling potential. On the other hand the line–shape of the resonance states is given by the Beutler–Fano function

$$F(q, \varepsilon) = \frac{(q + \varepsilon)^2}{1 + \varepsilon^2} \qquad (6.93)$$

Here q is called as the shape parameter and the reduced energy ε is given by

$$\varepsilon = \frac{E - E_R}{\Gamma/2}. \qquad (6.94)$$

Fano formalism consideres both bound and continuum regions and requires the wave functions of both regions. It is also possible to predict resonance states considering only bound state solutions. For example, time stability theory (Öksüz, 1976) is able to predict the energy values of resonance states considering only the discrete region wave functions from $\delta\sigma = min.$, where $\sigma =< H^2 > - < H >^2$. On the other hand, derivative method (Erkoç and Öksüz, 1979) also predicts the resonance states using bound state wave functions from $\frac{\partial}{\partial \lambda} < H >_\phi = min$. The resonance widths may also be calculated (Erkoç and Öksüz, 1979) from

$$\frac{2(E - E_R)}{\Gamma} \cong \sqrt{\frac{\sigma_\phi(E)}{\sigma_\phi(E_R)} - 1}$$

The $\sigma_\phi(E)$ values here may be obtained from the time stability spectrum (Öksüz, 1976), expressed by the eigenvalue equation

$$(H^2 - 2EH)|\phi > = (< H^2 > - 2E < H >)|\phi >$$

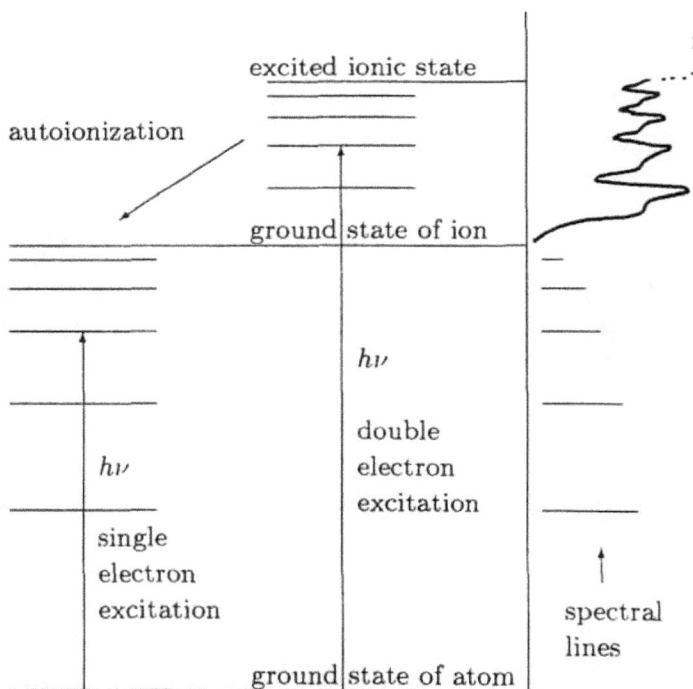

Fig. 6.6. Shematic representation of autoionization for many–electron atoms.

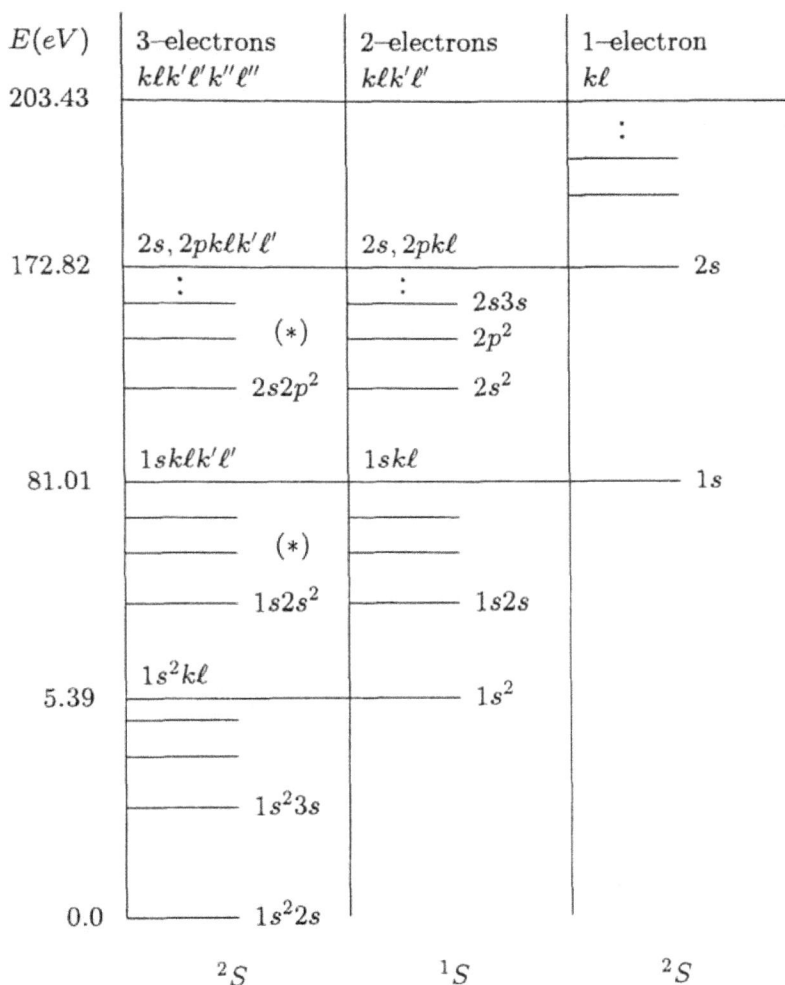

Fig. 6.7. Approximate energy level diagram for Li, Li^+, and Li^{++}. (∗) States lying in these regions are usually known as core–hole states, which are also called resonance or autoionizing states.

6.11   Screening and the orbital energies

The $2s$ and $2p$ levels, which are degenerate in a one–electron atom, are split when more than one electron is present. Compare the two lithium configurations $1s^2 2s$ and $1s^2 2p$. In both configurations, the electron in the $n = 2$ orbital is partially shielded from the nuclear attraction by the $1s$ electrons. The shielding is different for $2s$ and $2p$ orbitals. The radial distribution functions for the $2s$ and $2p$ electrons qualitatively look like

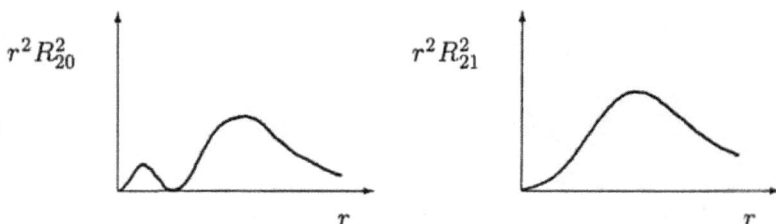

Fig. 6.8. Radial distribution functions for $2s$ and $2p$ orbitals.

The $2s$ distribution has a **hump** close to the nucleus, while the $2p$ distribution function is very small in this region. A $2s$ electron thus penetrates inside the $1s$ distribution to a greater extent than does a $2p$ electron. This results in less shielding of the $2s$ than the $2p$ electron by the inner–shell $1s$ electrons, so that the effective nuclear charge Z_e for a $2s$ electron is greater than that of a $2p$ electron; therefore the $2s$ electron with a larger Z_e is lower in energy (more stable) than the $2p$ electron (since $E \propto Z_e$).

$$(Z_e)_{2s} > (Z_e)_{2p} \quad \longrightarrow \quad E_{2s} < E_{2p}$$

A crude estimate of the $2s$, $2p$ splitting can be made by perturbation theory. For the Li atom in the $1s^2 2s$ state, we have

$$E_{Li}(1s^2 2s) = 2E_{1s} + E_{2s} + \, <V_{ee}>_{1s^2 2s} \qquad (6.95)$$

where the $e - e$ repulsion term is

$$<V_{ee}>_{1s^2 2s} = \int \psi_{Li}(1,2,3) \left(\frac{e^2}{r_{12}} + \frac{e^2}{r_{13}} + \frac{e^2}{r_{23}} \right) \psi_{Li}(1,2,3) dv_1 dv_2 dv_3$$
$$(6.96)$$

where $\psi_{Li}(1, 2, 3)$ represents the $1s^2 2s$ wave function.

The corresponding value for the $1s^2 2p$ state is

$$E_{Li}(1s^2 2p) = 2E_{1s} + E_{2p} + <V_{ee}>_{1s^2 2p} \qquad (6.97)$$

Since $E_{2s} = E_{2p}$, the energy difference between the Li atom in the $1s^2 2s$ and $1s^2 2p$ states is given by the difference in the $e - e$ repulsion terms $<V_{ee}>_{1s^2 2s}$ and $<V_{ee}>_{1s^2 2p}$. The perturbation theory result is $0.41\ eV$, while the experimental value is $1.85\ eV$. The magnitude of splitting increases rapidly with Z.

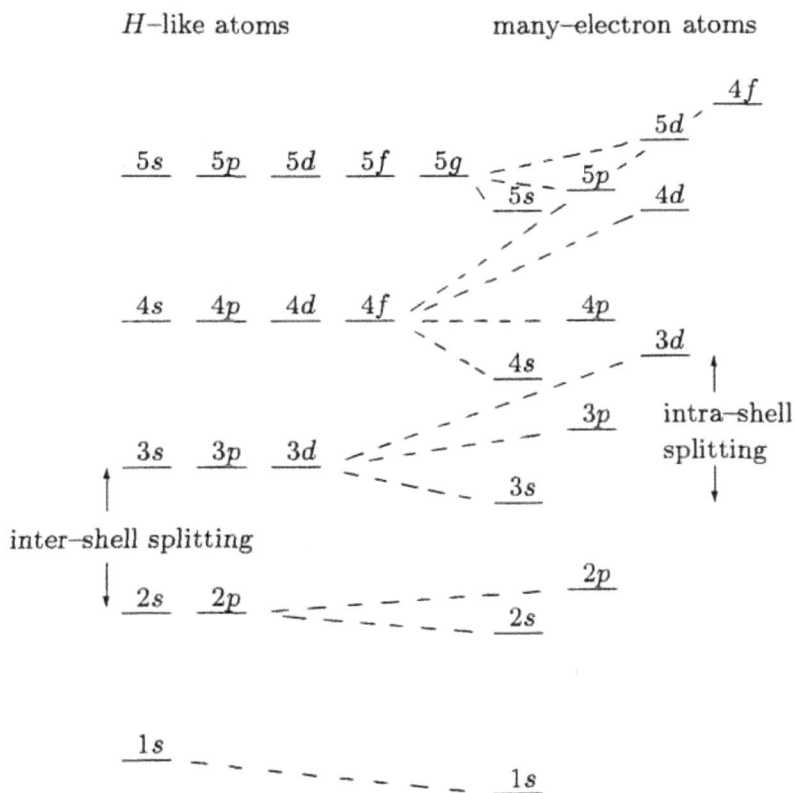

Fig. 6.9. Splitting of orbital degeneracy in many–electron atoms.

In some cases the intrashell splitting (same n, different ℓ) is larger than the intershell splitting (different n), so that an inversion of level order occurs. For example, the 4s level is lower than the 3d level in the diagram. This level ordering is common for neutral atoms. However, a detailed examination of spectra, careful calculation, or both, are required to establish the exact order in each particular atom or ion.

As Z increases, the electron–nuclear attraction becomes larger since the shielding by the added electrons is incomplete. Thus the inner–orbital levels decrease sharply in energy with increasing Z, while the energy levels of the more shielded outer orbitals decrease less sharply with Z.

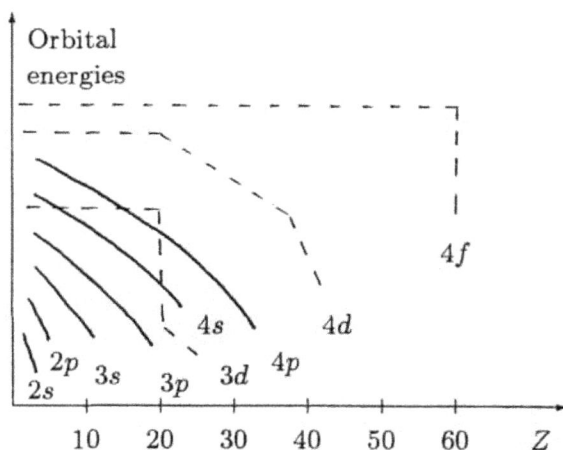

Fig. 6.10. Variation of orbital energies with atomic number.

6.12   The Aufbau principle and the periodic table

The usual description of atomic properties and their periodic classification is mainly concerned with the ground state. The atomic structure (electronic structure) is different for each possible state.

The most striking features of the periodic table are the following:

- The table consists of rows (periods) with 2, 8, 8, 18, 18, 32, and 32 elements.
- Elements in the same column (group) have similar chemical properties and atomic spectra.
- The first IP increases gradually across a period but drops sharply between periods.
- The long periods (4–7) contain sets of elements with similar chemical properties.

All of these features can be accounted for when the Pauli principle and the energy ordering of orbitals are used to obtain ground state electron configurations for the elements.

An empirical order of the orbitals (Aufbau principle) is

$$1s < 2s < 2p < 3s < 3p < 4s < 3d < 4p < 5s$$

$$5s < 4d < 5p < 6s < 4f < 5d < 6p < 7s$$

Table 6.1: Some exceptions (ground state electron configurations):

predicted	actually found	element
$ns^2(n-1)d^4$	$ns(n-1)d^5$	Cr , Mo
$ns^2(n-1)d^9$	$ns(n-1)d^{10}$	Cu , Ag , Au
$6s^2 4f$	$6s^2 5d$	La

A set of orbitals with the same value of n is called a **shell**.

Standard notation for the shells:

$$n = 1(K) \ , \ 2(L) \ , \ 3(M) \ , \ 4(N) \ , \ 5(O) \ , \ 6(P)$$

Maximum number of electrons in an orbital (a subshell): $2(2\ell + 1)$

Configuration for full subshells: $n\ell^{2(2\ell+1)}$

$$\ell \;=\; 0 \;,\; 1 \;,\; 2 \;,\; 3 \;,\; 4 \;,\; 5 \;,\; \ldots$$
$$ s^2 \;,\; p^6 \;,\; d^{10} \;,\; f^{14} \;,\; g^{18} \;,\; h^{22} \;,\; \ldots$$

Total number of electrons in a shell: $2n^2$

$$
\begin{array}{llllrlr}
n = 1 \,, & K & : & s^2 & : & 2 & : & 2 \\
n = 2 \,, & L & : & s^2 p^6 & : & 8 & : & 10 \\
n = 3 \,, & M & : & s^2 p^6 d^{10} & : & 18 & : & 28 \\
n = 4 \,, & N & : & s^2 p^6 d^{10} f^{14} & : & 32 & : & 60 \\
n = 5 \,, & O & : & s^2 p^6 d^{10} f^{14} g^{18} & : & 50 & : & 110 \\
n = 6 \,, & P & : & s^2 p^6 d^{10} f^{14} g^{18} h^{22} & : & 72 & : & 182
\end{array}
$$

Empirical order of the orbitals:

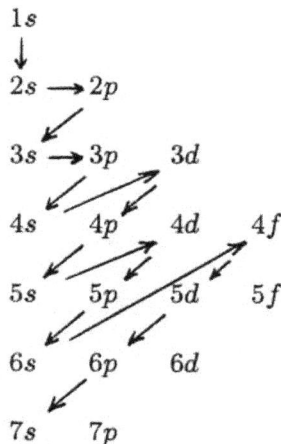

$$1s$$
$$\downarrow$$
$$2s \longrightarrow 2p$$
$$3s \longrightarrow 3p \qquad 3d$$
$$4s \qquad 4p \qquad 4d \qquad 4f$$
$$5s \qquad 5p \qquad 5d \qquad 5f$$
$$6s \qquad 6p \qquad 6d$$
$$7s \qquad 7p$$

Periodic table of the elements:

alkali metals (except *H*) noble gases (rare gases)

alkali earth metals

transition metals metallic earths

halogens

first transition series

second transition series

third transition series

		He	$2p^6$
	F	Ne	$3p^6$
		Ar	$4p^6$
		Kr	$5p^6$
		Xe	
		Rn	

	B	C	N	O		
	Al					

	Zn
	Cd
	Hg

1	*H*			
2	*Li*	*Be*		
$[Ne]3s$ 3	*Na*	*Mg*		
$[Ar]4s$ 4	*K*	*Ca*	*Sc*	*Zn*
$[Kr]5s$ 5	*Rb*		*Y*	*Cd*
6	*Cs*		*La*	*Hf*	*Hg*
7	*Fr*		*Ac*	

4*f* series | *Ce* | | | *Lu* | Lanthanides (rare earths)

5*f* series | *Th* | *U* | | *Lr* | Actinides

Periodic table of the elements:

1 H																	2 He
3 Li	4 Be											5 B	6 C	7 N	8 O	9 F	10 Ne
11 Na	12 Mg											13 Al	14 Si	15 P	16 Si	17 Cl	18 Ar
19 K	20 Ca	21 Sc	22 Ti	23 V	24 Cr	25 Mn	26 Fe	27 Co	28 Ni	29 Cu	30 Zn	31 Ga	32 Ge	33 As	34 Se	35 Br	36 Kr
37 Rb	38 Sr	39 Y	40 Zr	41 Nb	42 Mo	43 Tc	44 Ru	45 Rh	46 Pd	47 Ag	48 Cd	49 In	50 Sn	51 Sb	52 Te	53 I	54 Xe
55 Cs	56 Ba	57 La	72 Hf	73 Ta	74 W	75 Re	76 Os	77 Ir	78 Pt	79 Au	80 Hg	81 Tl	82 Pb	83 Bi	84 Po	85 At	86 Rn
87 Fr	88 Ra	89 Ac	104 Unq	105 Unp	106 Unh												

58 Ce	59 Pr	60 Nd	61 Pm	62 Sm	63 Eu	64 Gd	65 Tb	66 Dy	67 Ho	68 Er	69 Tm	70 Yb	71 Lu
90 Th	91 Pa	92 U	93 Np	94 Pu	95 Am	96 Cm	97 Bk	98 Cf	99 Es	100 Fm	101 Md	102 No	103 Lr

6.13   Vector model of the atom

The non–relativistic N–electron Hamiltonian (in $a.u.$) is given by

$$H_0 = -\frac{1}{2} \sum_{i=1}^{N} \nabla_i^2 - Z \sum_{i=1}^{N} \frac{1}{r_i} + \sum_{i<j} \frac{1}{r_{ij}} \qquad (6.98)$$

The relativistic N–electron Hamiltonian is

$$H = H_0 + H_{SL}^{(1)} + H_S^{(2)} + H_L^{(3)} + H_{SN}^{(4)} + H_{SI}^{(5)} + H_{IL}^{(6)} + H_{NQ}^{(7)} \qquad (6.99)$$

where the additional terms represent the following contributions
(1): Spin–orbit interaction
(2): Spin–spin interaction
(3): Orbit–orbit interaction
(4): Nuclear spin–nuclear spin interaction (ESR)
(5): Electron spin–nuclear spin interaction (NMR)
(6): Nuclear spin–orbit interaction
(7): Nuclear quadrupole interaction

Spin–orbit interaction ($L - S$ Coupling):

The physical meaning of $L - S$ coupling means that neither **L** nor **S** are not constants of motion separately, because they both precess around **J**, as shown in the figure.

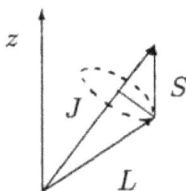

Fig. 6.11. Precession of L and S around J.

This precession is due to the torque which is caused by the interaction of orbital angular momentum and magnetic moment of the electron.

Energy of an electron with magnetic moment $\vec{\mu}$ in external magnetic field $\mathbf{B}$ is given by

$$E_{mag} = -\vec{\mu} \cdot \mathbf{B} \tag{6.100}$$

Magnetic moment of the electron is due to its spin

$$\vec{\mu} \rightarrow \vec{\mu}_s = -\frac{e}{mc}\mathbf{S} \tag{6.101}$$

Magnetic field of the nucleus on moving electron is given by

$$\mathbf{B} = \frac{Ze}{2mcr^3}\mathbf{L} \tag{6.102}$$

Thus, E_{mag} takes the form

$$E_{mag} = \frac{Ze^2}{2m^2c^2r^3}\mathbf{L} \cdot \mathbf{S} \tag{6.103}$$

in operator form

$$H_{SL} = \frac{Ze^2}{2m^2c^2r^3}\mathbf{L}_{op} \cdot \mathbf{S}_{op} = \xi(r)\mathbf{L}_{op} \cdot \mathbf{S}_{op} \tag{6.104}$$

For N–electron atom:

$$H_{SL} = \sum_{i=1}^{N} \xi(r_i)\left(L_{x_i}S_{x_i} + L_{y_i}S_{y_i} + L_{z_i}S_{z_i}\right) \tag{6.105}$$

Let us define the central–field Hamiltonian as

$$H_1 = \sum_{i=1}^{N} \left(-\frac{1}{2}\nabla_i^2 + V(r_i)\right) \tag{6.106}$$

where $V(r) = -Z/r$. Define the $e - e$ repulsion terms be represented as

$$H_{12} = \sum_{i<j}^{N} \frac{1}{r_{ij}} \tag{6.107}$$

Then the non–relativistic Hamiltonian becomes

$$H_0 = H_1 + H_{12} \tag{6.108}$$

The spin–orbit Hamiltonian is defined as

$$H = H_0 + H_{SL} = H_1 + H_{12} + H_{SL} \tag{6.109}$$

The orbital angular momentum and spin angular momentum operators L_z, L^2, S_z and S^2 commute with H_1 and H_{12}. Using the commutation relations

$$[L_\alpha, L_\beta] = i\varepsilon_{\alpha\beta\gamma}L_\gamma \tag{6.110}$$

and

$$[S_\alpha, S_\beta] = i\varepsilon_{\alpha\beta\gamma}S_\gamma \tag{6.111}$$

where $\varepsilon_{\alpha\beta\gamma}$ is the Levi–Civita tensor, and has the values $\varepsilon_{\alpha\beta\gamma} = +1$, $\varepsilon_{\alpha\gamma\beta} = -1$, $\varepsilon_{\alpha\beta\beta} = \varepsilon_{\beta\beta\gamma} = \varepsilon_{\alpha\alpha\gamma} = \varepsilon_{\alpha\gamma\gamma} = 0$, one can show that

$$[L_z, \mathbf{L_{op}} \cdot \mathbf{S_{op}}] = i(L_y S_x - L_x S_y) \tag{6.112}$$

$$[S_z, \mathbf{L_{op}} \cdot \mathbf{S_{op}}] = i(S_y L_x - S_x L_y)$$

these relations, Eq.(6.112), show that both L_z and S_z do not commute with H_{SL}; but their sum, i.e., $L_z + S_z = J_z$ commutes with H_{SL},

$$[(L_z + S_z), \mathbf{L_{op}} \cdot \mathbf{S_{op}}] = 0 \tag{6.113}$$

or

$$[J_z, H_{SL}] = 0 \tag{6.114}$$

It is also possible to show that

$$[J_\alpha^2, H_{SL}] = 0 \quad , \quad \alpha = x, y, z \tag{6.115}$$

$$[J^2, H_{SL}] = 0 \tag{6.116}$$

or

$$[J^2, H] = 0 \tag{6.117}$$

Thus, $\mathbf{J}$ is constant of motion, but $\mathbf{L}$ and $\mathbf{S}$ are not.

For small Z values ($Z \leq 40$) $H_{12} \gg H_{SL}$, therefore $H_0 = H_1 + H_{12}$ and H_{SL} is a small perturbation, and

$$\mathbf{J} = \mathbf{L} + \mathbf{S} \quad , \quad J = L + S, \; L + S - 1, \; ... \; , \; |L - S| \tag{6.118}$$

This type of coupling is called as **LS–coupling** or Russell–Saunders coupling.

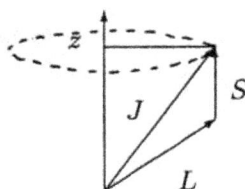

Fig. 6.12. LS–coupling model.

In LS–coupling

$$J^2 = (\mathbf{L_{op}} + \mathbf{S_{op}}) \cdot (\mathbf{L_{op}} + \mathbf{S_{op}}) = L^2 + S^2 + 2\mathbf{L_{op}} \cdot \mathbf{S_{op}} \qquad (6.119)$$

Therefore

$$\mathbf{L_{op}} \cdot \mathbf{S_{op}} = \frac{1}{2}(J^2 - L^2 - S^2) \qquad (6.120)$$

For large Z values $(Z > 40)$ $H_{SL} \gg H_{12}$, therefore $H_0 = H_1 + H_{SL}$ and H_{12} is a small perturbation. In this case L^2 and S^2 do not commute with H_0, but $J_z = L_z + S_z$ commutes with H_0, so

$$j_i = \ell_i + s_i \ , \ \ell_i + s_i - 1 \ , \ ... \ , \ |\ell_i - s_i| \qquad (6.121)$$

and

$$\mathbf{J} = \sum_{i=1}^{N} \mathbf{j}_i$$

This means that the orbital angular momentum and spin angular momentum of an electron couples strongly. This type of coupling is called as **jj–coupling**.

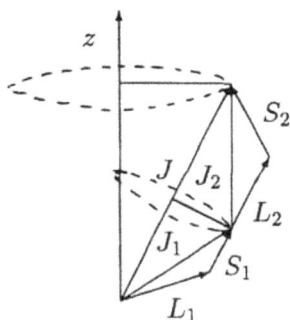

Fig. 6.13. jj–coupling model.

$LS - coupling$	$jj - coupling$
(H_{12}) : $\mathbf{L} = \sum_i \boldsymbol{\ell}_i$, $\mathbf{S} = \sum_i \mathbf{s}_i$	(H_{SL}) : $\mathbf{j}_i = \boldsymbol{\ell}_i + \mathbf{s}_i$
(H_{SL}) : $\mathbf{J} = \mathbf{L} + \mathbf{S}$	(H_{12}) : $\mathbf{J} = \sum_i \mathbf{j}_i$
$(H_{12} \gg H_{SL})$	$(H_{SL} \gg H_{12})$
$Z \leq 40$	$Z > 40$

6.14   Term symbols for lighter atoms

Lighter atoms belong to LS–coupling (or Russell–Saunders coupling) scheme. For a group of k electrons, the possible values of the total spin angular momentum are

$$S = \frac{k}{2} , \; \frac{k}{2} - 1 , \; \frac{k}{2} - 2 , \; ...., \; , \; 0 \quad (k \;\; even) \tag{6.122}$$

$$S = \frac{k}{2} , \; \frac{k}{2} - 1 , \; \frac{k}{2} - 2 , \; ...., \; , \; \frac{1}{2} \quad (k \;\; odd)$$

The possible values of the total orbital angular momentum are

$$L = \ell_1 + \ell_2 + ... + \ell_k , \; \ell_1 + \ell_2 + ... + \ell_k - 1 , \; \ell_1 + \ell_2 + ... + \ell_k - 2 , \; ... \tag{6.123}$$

For example, for three p electrons we have the possible L and S values:

$$\ell_1 = \ell_2 = \ell_3 = 1$$

$$s_1 = s_2 = s_3 = 1/2$$

$$L = 3, 2, 1, 0$$

$$S = 3/2, 1/2$$

For one f electron and two p electrons we have

$$\ell_1 = 3 \ , \ \ell_2 = \ell_3 = 1 \ ; \quad s_1 = s_2 = s_3 = \tfrac{1}{2}$$

$$L = 5, 4, 3, 2, 1 \ ; \quad S = \tfrac{3}{2}, \tfrac{1}{2}$$

The possible total angular momentum quantum numbers are obtained by writing

$$J = L + S \ , \ L + S - 1 \ , \ \dots \ , \ |L - S|.$$

For example, for $L = 3$ and $S = \tfrac{3}{2}$, we have $J = \tfrac{9}{2} \ , \ \tfrac{7}{2} \ , \ \tfrac{5}{2} \ , \ \tfrac{3}{2}$.

For each set of values, L, S, and J, the term symbol is written as in the one–electron case

$$^{2S+1}\mathbf{L}_J$$

The term symbol for an atom with $L = 3$, $S = 3/2$, $J = 5/2$ is $^4F_{5/2}$.

All possible values of S, L, and J values are the result of vector addition, they start from a maximum value and go in integer steps down to a minimum value.

In the above examples we have ignored the Pauli principle. As long as the n and/or ℓ quantum numbers of all the electrons in the open shell set are different, all of the possible terms can occur; that is, each combination of L and S is possible; such electrons are called **non–equivalent**

electrons. However, if some of the electrons have the same n and ℓ quantum numbers, the Pauli principle restricts the m_ℓ and m_s quantum numbers; a limit is thereby placed on the different terms; this type of electrons are called **equivalent electrons.**

Consider two p electrons:

$$\ell_1 = 1 \ , \ m_{\ell_1} = 1, 0, -1 \ , \ s_1 = 1/2 \ , \ m_{s_1} = 1/2, -1/2$$

$$\ell_2 = 1 \ , \ m_{\ell_2} = 1, 0, -1 \ , \ s_2 = 1/2 \ , \ m_{s_2} = 1/2, -1/2$$

$$M_L = m_{\ell_1} + m_{\ell_2} \ , \ \mathbf{L} = \ell_1 + \ell_2 \ ; \quad M_S = m_{s_1} + m_{s_2} \ , \ \mathbf{S} = s_1 + s_2$$

$$M_J = M_L + M_S \ , \ \mathbf{J} = \mathbf{L} + \mathbf{S} \ \text{ (in the } LS\text{–coupling scheme)}$$

For non–equivalent electrons: $npn'p$

$$L = 2, 1, 0 \ ; \ S = 1, 0$$

$$J_{LS=21} = 3, 2, 1 \quad : \quad {}^3D_3 \ {}^3D_2 \ {}^3D_1$$

$$J_{11} = 2, 1, 0 \quad : \quad {}^3P_2 \ {}^3P_1 \ {}^3P_0$$

$$J_{01} = 1 \quad : \quad {}^3S_1 \ ; \quad J_{20} = 2 \quad : \quad {}^1D_2$$

$$J_{10} = 1 \quad : \quad {}^1P_1 \ ; \quad J_{00} = 0 \quad : \quad {}^1S_0$$

For equivalent electrons: np^2

They must at least differ in their values of m_ℓ or m_s.

If $\ell_1 = \ell_2 = 1$ and $m_{\ell_1} = m_{\ell_2} = 1$ ($M_L = 2 \to D$; $L \to D$) then $m_{s_1} \neq m_{s_2}$ so $S = 0$ is possible only for D states, eliminate ${}^3D$ states.

If $\ell_1 = \ell_2 = 1$ and $m_{\ell_1} = m_{\ell_2} = 0$ ($M_L = 0 \to S$; $L \to S$) then $m_{s_1} \neq m_{s_2}$ so $S = 0$ is possible only for S states, eliminate ${}^3S$ states.

If $\ell_1 = \ell_2 = 1$ and $m_{\ell_1} = 1$, $m_{\ell_2} = 0$ or $m_{\ell_1} = 0$, $m_{\ell_2} = 1$ $(M_L = 1 \rightarrow P$; $L \rightarrow P)$ then $m_{s_1} = m_{s_2}$ so $S = 1$ is possible, eliminate ^1P states.

Therefore, the remaining states are $^3P_{2,1,0}$, $^1D_2$, $^1S_0$.

If the configuration contains both equivalent and non–equivalent electrons, one first finds the terms for the equivalent electrons and then couples in the non–equivalent electrons.

Terms for sss configuration:

For non–equivalent electrons: $\ell_1 = \ell_2 = \ell_3 = 0$, $s_1 = s_2 = s_3 = 1/2$; $L = 0 \rightarrow S$, $S = 3/2, 1/2$, $2S + 1 = 4, 2$. Therefore, the possible terms are $^2S_{1/2}$, $^4S_{3/2}$.

For both equivalent and non–equivalent electrons:

$ss \rightarrow L = 0$, $S = 1, 0$

$L = 0$, $S = 0 \rightarrow J = 0 \rightarrow^1 S_0$; $L = 0$, $S = 1 \rightarrow J = 1 \rightarrow^3 S_1$

$ss(\,^1S_0) + s \rightarrow L = 0$, $S = 1/2 \rightarrow J = 1/2 \rightarrow^2 S_{1/2}$

$ss(\,^3S_0) + s \rightarrow L = 0$, $S = 3/2, 1/2 \rightarrow J = 3/2, 1/2 \rightarrow^2 S_{1/2}$, $^4S_{3/2}$

Therefore, $(sss) \rightarrow$ ^2S ^2S ^4S .

Table 6.2: Terms of non–equivalent electrons:

Electron configuration	Terms
ss	$^1S\ ^3S$
sp	$^1P\ ^3P$
sd	$^1D\ ^3D$
pp	$^1S\ ^1P\ ^1D\ ^3S\ ^3P\ ^3D$
pd	$^1P\ ^1D\ ^1F\ ^3P\ ^3D\ ^3F$
dd	$^1S\ ^1P\ ^1D\ ^1F\ ^1G\ ^3S\ ^3P\ ^3D\ ^3F\ ^3G$
sss	$^2S\ ^2S\ ^4S$
ssp	$^2P\ ^2P\ ^4P$
ssd	$^2D\ ^2D\ ^4D$
spp	$^2S\ ^2P\ ^2D\ ^2S\ ^2P\ ^2D\ ^4S\ ^4P\ ^4D$
spd	$^2P\ ^2D\ ^2F\ ^2P\ ^2D\ ^2F\ ^4P\ ^4D\ ^4F$

Table 6.3: Terms of equivalent electrons:

Electron configuuration	Terms
ns^2	^1S
$np,\ np^5$	^2P
$np^2,\ np^4$	$^1S\ ^1D\ ^3P$
np^3	$^2P\ ^2D\ ^4S$
$nd,\ nd^9$	^2D
$nd^2,\ nd^8$	$^1S\ ^1D\ ^1G\ ^3P\ ^3F$

Table 6.4: Possible multiplicities for various number of electrons:

# of e^-	Possible multiplicities $(2S + 1)$
1	doublets(2)
2	singlets(1) , triplets(3)
3	doublets(2) , quartets(4)
4	singlets(1) , triplets(3) , quintets(5)
5	doublets(2) , quartets(4) , sextets(6)
6	singlets(1) , triplets(3) , quintets(5) , septets(7)
7	doublets(2) , quartets(4) , sextets(6) , octets(8)
8	singlets(1) , triplets(3) , quintets(5) , septets(7) , nonets(9)

Another approach to determine the term values of electronic configurations of atoms in LS–coupling scheme: (Spin–orbit coupling is negligible and there is no external field)

First look at the possible wave functions for the given configuration. If the configuration is in a closed–shell structure, then a single wave function is enough to represent the system (we mean single determinantal wave function). Otherwise there are more than one possibilities for the determinantal wave functions.

For a given ℓ, the maximum number of electrons in an orbital is $N_0 = 2(2\ell + 1)$; if there are say $N_1(\leq N_0)$ electrons in a given ℓ, the possibility of constructing determinantal wave functions is

$$N = \frac{N_0!}{N_1!(N_0 - N_1)!} \tag{6.124}$$

For example, for p^2 configuration, ($N_0 = 6$, $N_1 = 2$; $N = 6!/(2!4!) = 15$) there are 15 possible wave functions (or terms). $\ell_1 = \ell_2 = 1$ and $s_1 = s_2 = 1/2$. The possible values for L ($= 2, 1, 0$) and for S ($= 1, 0$). Therefore the possible terms are ^3S , ^3P , ^3D , ^1S , ^1P , ^1D. However, according to the Pauli principle some of these terms are eliminated.

We construct an M_L table: ($M_L = \sum_i m_{\ell_i}$)

Table 6.5: M_L table for p^2 configuration:

m_{ℓ_2} \ m_{ℓ_1}	1	0	-1
1	2	1	0
0	1	0	-1
-1	0	-1	-2

1. group $\rightarrow$ 2   1   0 -1 -2 $(L=2)$ D
2. group $\rightarrow$ 1   0 -1 $(L=1)$ P
3. group $\rightarrow$ 0 $(L=0)$ S

We look at the diagonal elements of the M_L table, ($|M_L| = 2, 0$) for these elements two electrons have the same (n, ℓ, m_ℓ), therefore to be able to obey Pauli principle these electrons must have opposite spins, in other words these terms cannot be triplet, eliminate ($M_L = 2 \rightarrow L = 2 \rightarrow {}^3D$; $M_L = 0 \rightarrow L = 0 \rightarrow {}^3S$) terms. Since $M_L = 1$ (or $L = 1$) is off–diagonal, the corresponding term could be either spin paired ($\uparrow\downarrow$) (singlet) or spin unpaired ($\uparrow\uparrow$, or $\downarrow\downarrow$) (triplet). The system chooses the lowest energy term (triplet), so ${}^3P$ is possible.

Consider nd^2 configuration:

Table 6.6: M_L table for d^2 configuration:

m_{ℓ_2} \ m_{ℓ_1}	2	1	0	-1	-2
2	4	3	2	1	0
1	3	2	1	0	-1
0	2	1	0	-1	-2
-1	1	0	-1	-2	-3
-2	0	-1	-2	-3	-4

$$
\begin{array}{llrrrrrrrrl}
1.\ group & \to & 4 & 3 & 2 & 1 & 0 & -1 & -2 & -3 & -4 & (L=4)\ G \\
2.\ group & \to & & 3 & 2 & 1 & 0 & -1 & -2 & -3 & & (L=3)\ F \\
3.\ group & \to & & & 2 & 1 & 0 & -1 & -2 & & & (L=2)\ D \\
4.\ group & \to & & & & 1 & 0 & -1 & & & & (L=1)\ P \\
5.\ group & \to & & & & & 0 & & & & & (L=0)\ S \\
\end{array}
$$

On the diagonal: $|M_L| = 4, 2, 0$. Therefore $L = 0, 2, 4$ must be singlet (^1S , ^1D , ^1G), the remaining terms (in the off–diagonal) $L = 1, 3$ must be triplet (^3P , ^3F).

6.15   Ground state terms, Hund's rules

Once the terms are known, it is necessary to find their relative energy in order to choose which term characterizes the ground state of the atom. This can be done by a set of simple rules, called Hund's rules. (These empirical rules are usually valid for equivalent electrons, but they are often also applicable to non–equivalent electrons):

a) The terms are ordered according to their S values, the term with maximum S being most stable and the stability decreasing with decreasing S. Thus, the ground state has maximum spin multiplicity.

b) For a given value of S (given spin multiplicity), the state with maximum L is most stable.

c) For a given S and L, the minimum J value is most stable if there is a single open subshell that is less than half–full and the maximum J is most stable if the subshell is more than half–full.

Rules (a) and (b) arise from the $e - e$ (electrostatic) interaction between the electrons, while rule (c) is a consequence of the spin–orbit (magnetic) interaction.

One can also interpret these rules as follows: When the spins of more than one electron from a given subshell are coupled to total spin S, then the state with the largest value of S is energetically lowest, because it feels the effects of the short ranged electron–electron repulsion least due to the symmetry properties of the spatial part of the wave function. This is Hund's first rule. For a given value of S the electrons can couple to different values L of the total orbital angular momentum. Amongst these states, the effect of the short ranged repulsion is least in the states with the largest values of L. Of all states with the same value of S, the state with the maximum value of L is hence the energetically lowest. This is Hund's second rule.

For example, the configurations np^2 and np^4 both give rise to the terms $^3P_{2,1,0}$ $^1D_2$ $^1S_0$; for np^2 the ground state term is $^3P_0$, while for np^4 it is $^3P_2$.

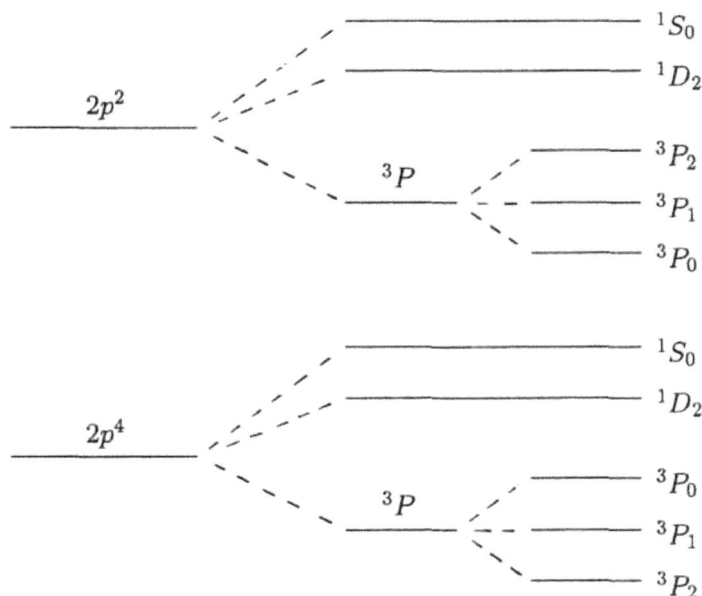

Fig. 6.14. Order of the energy levels for p^2 and p^4 configurations.

The splitting and the order of the single degenerate energy level of a particular configuration, say the $ndn'p$ configuration of an atom with two optically active electrons, due to the residual Coulomb and spin–orbit perturbations may be shown as follows:

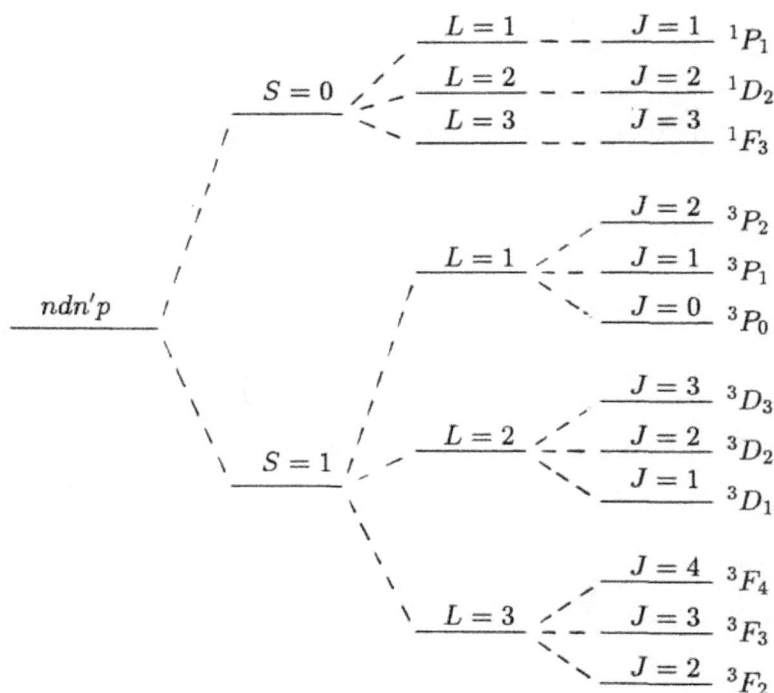

Fig. 6.15. Splitting of energy levels in LS–coupling.

LS–coupling $\left(^{2S+1}L_J\right)$ jj–coupling $(j_1, j_2)_J$

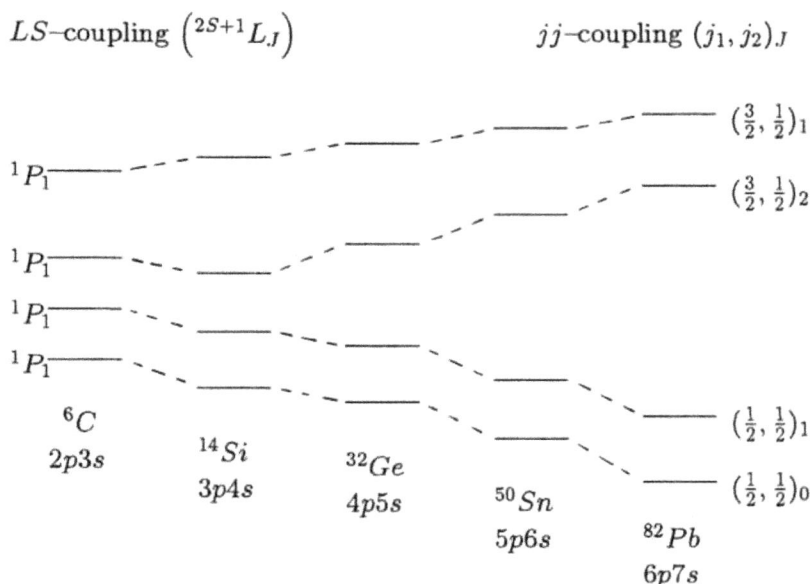

$(\frac{3}{2}, \frac{1}{2})_1$

$^1P_1$

$(\frac{3}{2}, \frac{1}{2})_2$

$^1P_1$

$^1P_1$

$^1P_1$

^6C

$2p3s$

^{14}Si

$3p4s$

^{32}Ge

$4p5s$

$(\frac{1}{2}, \frac{1}{2})_1$

^{50}Sn

$5p6s$

^{82}Pb

$6p7s$

$(\frac{1}{2}, \frac{1}{2})_0$

Fig. 6.16. Correlation diagram for LS– and jj–coupling schemes for the electron configuration $np(n+1)s$.

LS–coupling:

$\ell_1 = 1$, $s_1 = 1/2$; $\ell_2 = 0$, $s_2 = 1/2$ $\rightarrow$ $L = 1$, $S = 1, 0$

$L = 1$, $S = 1 \rightarrow {}^3P_{2,1,0}$; $L = 1$, $S = 0 \rightarrow {}^1P_1$

jj–coupling:

$\ell_1 = 1$, $s_1 = 1/2$ $\rightarrow$ $j_1 = 3/2, 1/2$; $\ell_2 = 0$, $s_2 = 1/2$ $\rightarrow$ $j_2 = 1/2$

(j_1, j_2):

$(3/2, 1/2)$ $\rightarrow$ $J = 2, 1$ $\rightarrow$ $(3/2, 1/2)_{1,2}$

$(1/2, 1/2)$ $\rightarrow$ $J = 1, 0$ $\rightarrow$ $(1/2, 1/2)_{1,0}$

6.16   Ionization potentials and electron affinities

The ionization potential (actually the first IP) is defined as the energy required for the process

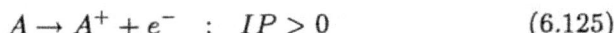

$$A \to A^+ + e^- \quad : \quad IP > 0 \qquad (6.125)$$

The electron affinity (actually the first EA) is defined as the energy released in the process

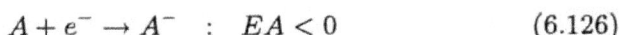

$$A + e^- \to A^- \quad : \quad EA < 0 \qquad (6.126)$$

($|IP| > |EA|$)

Plot of IP_1 of the elements as a function of atomic number Z looks as follows:

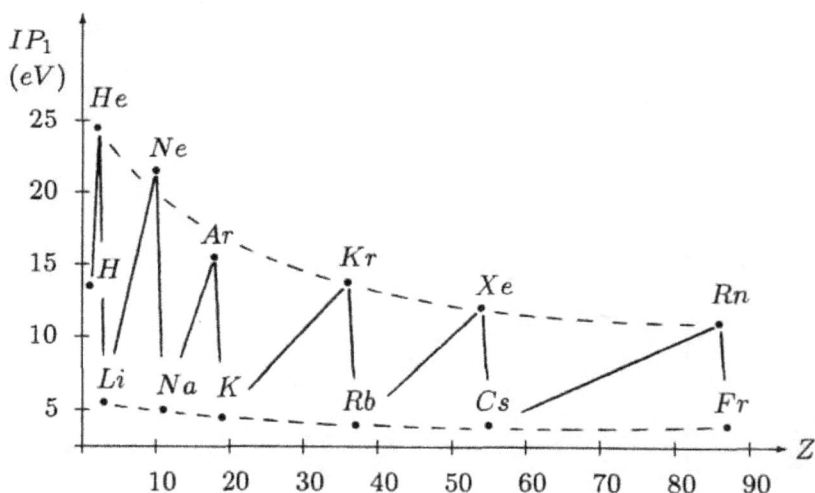

Fig. 6.17. Ionization potentials of noble gases and alkalis.

IP_1 is the energy required to remove one of the outermost electrons when the atom is in the ground state. EA_1 is the energy gained when an electron with no kinetic energy is added to the isolated atom.

The electron affinities follow trends similar to those for the IP.

In the process

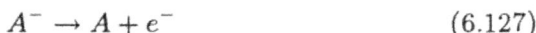

$$A^- \to A + e^- \tag{6.127}$$

the EA of A is the IP of A^-.

IP_1 may be approximated by

$$IP_1 \cong \frac{Z_e^2}{n^2} \left(\frac{e^2}{2a_0} \right) \tag{6.128}$$

The IP vs. Z diagram can be interpreted by considering this approximate expression for IP_1.

Another related quantity is the electronegativity (χ), which is a measure of the tendency of the atom to form ionic compounds. Mulliken's definition of electronegativity is

$$\chi = \frac{1}{2}(IP + EA) > 0 \tag{6.129}$$

6.17 Atomic radii

The radius of an atom is a quantity that is not exactly defined, its magnitude depending upon the method of measurement. An approximate equation for the effective radius of the electron cloud may be written as

$$r_{eff} \cong \frac{n^2}{Z_e} a_0 \tag{6.130}$$

The radial distribution function for a many–electron atom is given by the sum of the distribution functions of the occupied orbitals. For example for $He(1s^2)$

$$P_{He}(r) = 4\pi r^2 [P_1(r) + P_2(r)] \tag{6.131}$$

For $Ar(1s^2 2s^2 2p^6 3s^2 3p^6)$

$$P_{Ar}(r) = 4\pi r^2 \sum_i P_i(r) \tag{6.132}$$

Wave functions for many–electron atoms are usually calculated by Hartree–Fock or Self–Consistent–Field method. These methods will be described in Chapter 8.

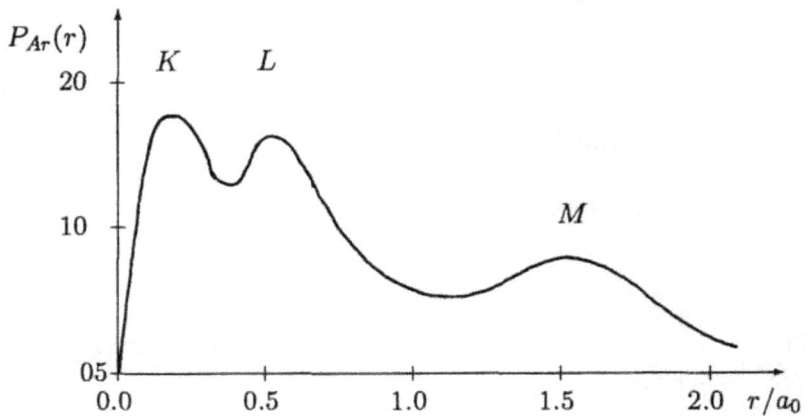

Fig. 6.18. Radial distribution function for Ar.

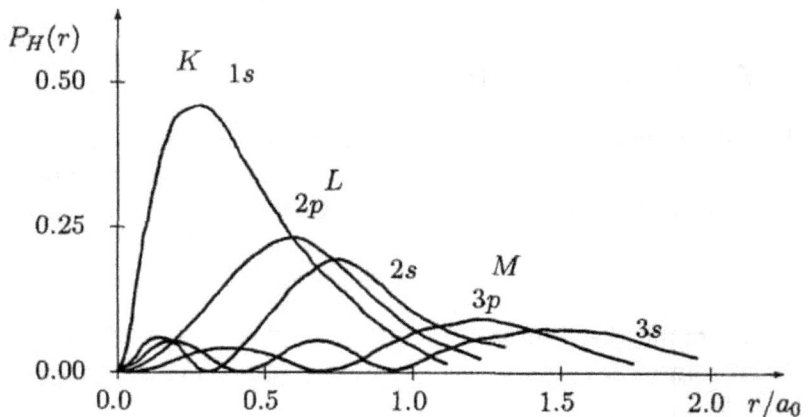

Fig. 6.19. Radial distribution functions for hydrogenic orbitals.

It is seen in these diagrams that there is a sharp separation of the individual electron shells. The last filled shell (the M shell) completely dominates the outer part of the radial distribution curve. This suggests that its form should determine the size of the atom.

6.18   Inner–shell transitions

The spectra of many–electron atoms can result from transitions of either inner–shell or outer–shell electrons.

The inner–shell electrons have relatively widely spaced energy levels, so that the radiation corresponding to transitions between them has wavelengths on the order of 0.1 to 100 $\mathring{A}$ (x–rays) and energies of 10^2 to 10^5 eV.

A common method of producing x–rays is to bombard a piece of metal with electrons of sufficient energy to knock an inner–shell electron out of a metal atom. The remaining electrons in the atom rearrange themselves by falling into the **holes** created and emit x–radiation in accordance with the Bohr frequency rule. Since the inner–shell energies change from metal to metal, the wavelengths of the x–rays vary with the target substance. Variation of x–ray spectra with Z have no simple periodicity.

H. Moseley (in 1913) found that a plot of the square root of the frequencies of the most energetic lines versus atomic number is nearly linear. A convenient empirical formula is **Moseley's law**.

$$\left(\frac{\bar{\nu}}{R}\right)^{1/2} = \left(\frac{1}{n_2^2} - \frac{1}{n_1^2}\right)^{1/2}(Z - \sigma) \qquad (6.133)$$

or

$$\frac{\bar{\nu}}{R} = \left(\frac{1}{n_2^2} - \frac{1}{n_1^2}\right)(Z - \sigma)^2 = \left(\frac{Z-\sigma}{n_2}\right)^2 - \left(\frac{Z-\sigma}{n_1}\right)^2$$

where R is the Rydberg constant and Z is the atomic number. The empirical constant σ can be interpreted as a **screening constant**; that is, its value is the amount of nuclear charge screened or shielded from the electron involved in the transition by the other electrons in the atom.

For the transitions ($n_1 = 2$, $n_2 = 1$) of each element, Eq.(6.133) gives a good fit to experimental data if σ is assigned the value 1.13 .

A more accurate formula, including relativistic effects (i.e., spin–orbit coupling) is

$$\frac{\bar{\nu}}{R} = \left(\frac{Z - \sigma_2}{n_2}\right)^2 - \left(\frac{Z - \sigma_1}{n_1}\right)^2 + \alpha^2 \left[\frac{(Z - \delta_2)^4}{n_2^3(\ell_2 + 1)} - \frac{(Z - \delta_1)^4}{n_1^3(\ell_1 + 1)}\right] + \ldots$$

(6.134)

where α is the dimensionless **fine–structure constant**

$$\alpha = \frac{e^2}{\hbar c} = 7.29720 \times 10^{-3} \cong \frac{1}{137}$$

(6.135)

In Eq.(6.134), the non–relativistic screening constant σ depends upon the configuration of the relevant shell and increases with increasing Z. The value of the relativistic screening constant δ also depends upon the shell configuration, but is independent of Z.

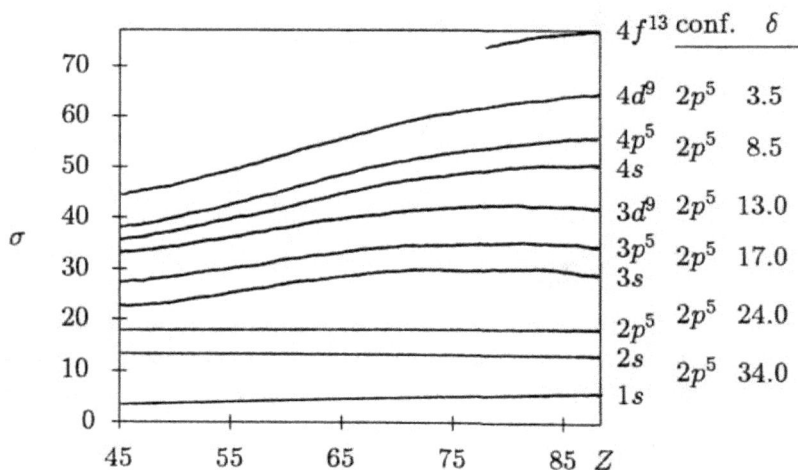

Fig. 6.20. Variation of σ versus Z and configuration.

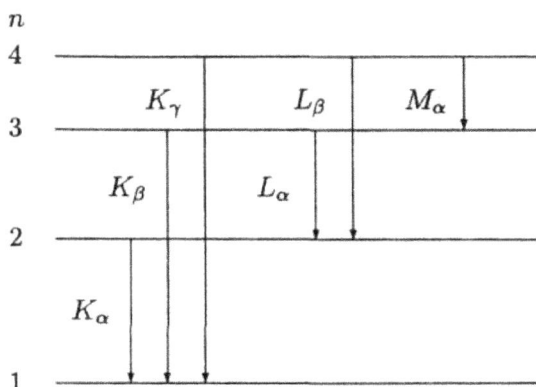

Fig. 6.21. The designation of X–ray lines.

Although x–ray spectra involve inner–shell electron of high energies, the emitted wavelengths are altered by the distribution of the outer–shell electrons.

Fine structure constant:

$$\alpha = \frac{e^2}{\hbar c} \xrightarrow{\text{in } SI} \frac{e^2}{(4\pi\varepsilon_0)\hbar c}$$

$$\alpha = \frac{(1.60219 \times 10^{-19})^2 \times 8.98755 \times 10^9}{1.05459 \times 10^{-34} \times 2.99792 \times 10^8}$$

$$\alpha = 7.2973582 \times 10^{-3} \quad , \quad \frac{1}{\alpha} = 137.035893$$

$$\alpha = \frac{e^2}{4\pi\varepsilon_0 \hbar c} = \frac{e^2}{2\varepsilon hc} = \frac{e^2 c\mu_0}{2h}$$

$$\alpha \cong \frac{(1.6 \times 10^{-19})^2 \times 3 \times 10^8 \times 4\pi \times 10^{-7}}{2 \times 6.626 \times 10^{-34}} = 7.283 \times 10^{-3} \cong \frac{1}{137}$$

6.19   Outer–shell transitions

The spectra arising from transitions of outer–shell electrons can be analyzed in terms of the electron configurations corresponding to the initial
and final states. For many–electron atoms, the simplest spectra are
those of the alkali metals, in which all the electrons but one occupy
closed inner shells. The configuration of these inner–shell electrons does
not change, so that the transitions of the single outer–shell electron give
rise to a spectrum which is similar to that of hydrogen.

The allowed transitions between the various levels are determined by
the selection rules (for electric dipole transitions):

$$\Delta L = \pm 1 \ , \quad \Delta S = 0 \ , \quad \Delta J = \pm 1 \ , \quad \Delta M_J = 0, \pm 1 \qquad (6.136)$$

The energy levels of alkali atoms can be represented by the empirical
formula

$$E_{n\ell} = -\frac{1}{2}\frac{1}{(n - \mu_{n\ell})^2} \quad a.u. \qquad (6.137)$$

The quantities $\mu_{n\ell}$ are known as **quantum defects**, and the quantity
$n^* = n - \mu_{n\ell}$ is defined as the effective principal quantum number

$$n^* \cong n - \alpha(\ell) - \frac{\beta(\ell)}{n^2} \qquad (6.138)$$

where α and β are a function of ℓ. For example, for Li, $\alpha(1) = 0.040$,
$\beta(1) = 0.024$.

The emission spectra of positive ions with one valence electron outside an inert core can also be observed in a spark discharge. The energy
levels of such a sequence can be written as

$$E_{n\ell} = -\frac{1}{2}\frac{\bar{Z}^2}{[n - \alpha(\ell)]^2} \quad a.u. \qquad (6.139)$$

where $\bar{Z} = Z - N + 1$, Z being the nuclear charge and N the number of
electrons.

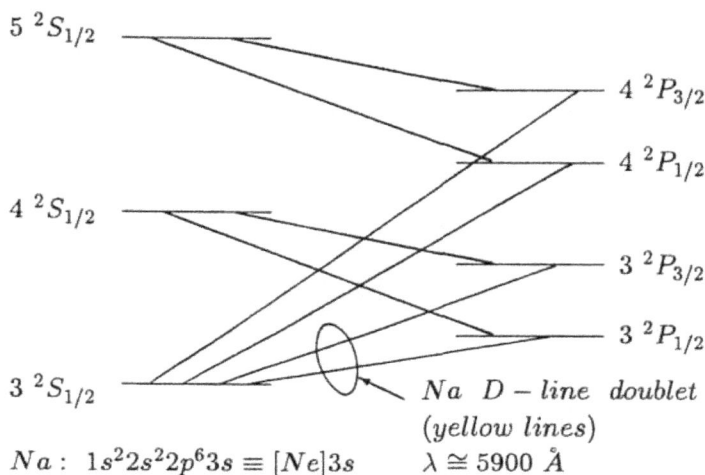

Fig. 6.22. Transitions among lowest terms of the Na atom.

6.20 Worked examples

Example - 6.1 :

The first manifold of excited states for the He atom is associated with the electronic transitions $1s^2 \rightarrow 1s2s$, $1s2p$. Set up simple product functions using hydrogen–like orbitals to represent the eight excited states, which are degenerate for noninteracting electrons, and use perturbation theory to derive expressions for the energies of these states.

Solution :

The product functions which represent the eight ways that the two electrons can be assigned to the $1s$, $2s$ and $2p$ orbitals in the singly excited states are:

$$u_1 = 1s(1)2s(2) \qquad u_2 = 1s(2)2s(1) \qquad u_3 = 1s(1)2p_x(2)$$
$$u_4 = 1s(2)2p_x(1) \qquad u_5 = 1s(1)2p_y(2) \qquad u_6 = 1s(2)2p_y(1)$$
$$u_7 = 1s(1)2p_z(2) \qquad u_8 = 1s(2)2p_z(1)$$

The Hamiltonian for He is

$$H = -\nabla_1^2 - 4/r_1 - \nabla_2^2 - 4/r_2 + 2/r_{12} = H_0 + H'$$

and we find that

$$H_0 u_i(1,2) = H_0 \psi_{n_1 \ell_1 m_{\ell_1}}(1) \psi_{n_2 \ell_2 m_{\ell_2}}(2) = -4(1/n_1^2 + 1/n_2^2) u_i = -5 u_i$$

since $n_1 = 1$ and $n_2 = 2$ for the first excited states.

Degenerate perturbation theory requires a secular equation of the form

$det|H'_{ij} - E'_n S_{ij}| = 0$. The required Coulomb (J) and exchange (K) integrals are:

$$
\begin{aligned}
J_s &= <1s(1)2s(2)|2/r_{12}|1s(1)2s(2)> \\
J_p &= <1s(1)2p(2)|2/r_{12}|1s(1)2p(2)> \\
K_s &= <1s(1)2s(2)|2/r_{12}|1s(2)2s(1)> \\
K_p &= <1s(1)2p(2)|2/r_{12}|1s(2)2p(1)>
\end{aligned}
$$

where p represents p_x, p_y or p_z. All of the other integrals vanish because of symmetry. For example,

$$H'_{23} = <u_2|2/r_{12}|u_3> = <2s(1)1s(2)|2/r_{12}|1s(1)2p_x(2)> = 0$$

since the integrand is an odd function of x. The secular determinant is:

$$
\begin{vmatrix}
J'_s & K_s & & & & & & \\
K_s & J'_s & & & & & & \\
& & J'_p & K_p & & & & \\
& & K_p & J'_p & & & & \\
& & & & J'_p & K_p & & \\
& & & & K_p & J'_p & & \\
& & & & & & J'_p & K_p \\
& & & & & & K_p & J'_p
\end{vmatrix} = 0
$$

where $J'_s = J_s - E'$ and $J'_p = J_p - E'$. The J_s, K_s block gives $(J_s - E')^2 - K_s^2 = 0$; $E' = J_s \pm K_s$ and the associated eigenfunctions are found to be
$\psi_{1,2} = (1/\sqrt{2})[1s(1)2s(2) \pm 1s(2)2s(1)]$.
The second diagonal block gives the roots $E' = J_p \pm K_p$ with the associated eigenfunctions
$\psi_{3,4} = (1/\sqrt{2})[1s(1)2p_x(2) \pm 1s(2)2p_x(1)]$.
Thus, this procedure leads to 8 states, 3 pairs of which are degenerate. It turns out that $J_p > J_s$ and all of the J'_s and K'_s are positive. The following diagram summarizes the calculation.

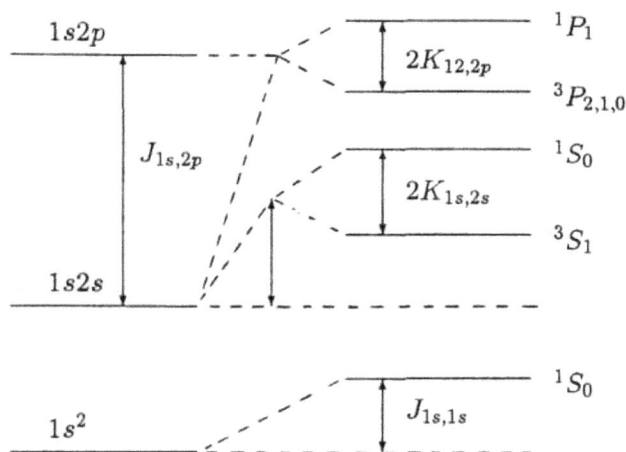

Example - 6.2 :

In the Coulomb approximation for the alkali atoms (Li, Na,...) the term energies are determined from the hydrogen–like formula $E_{n\ell} = E_{core} - Z_{net}^2/2(n^*)^2$, where Z_{net} and E_{core} are the net charge and energy, respectively, of the ion core. E_{core} is assumed to be constant for all atoms in an isoelectronic series, n^* is an effective principal quantum number, $n^* = n - \delta_\ell$, where δ_ℓ is the quantum defect or Rydberg correction. The ionization potentials of the isoelectronic Li series are 5.390 $eV(Li)$, 18.206 $eV(Be^+)$, 37.920 $eV(B^{2+})$, and 64.476 $eV(C^{3+})$.
a) Use these ionization potentials to determine the quantum defects δ_s

for Li, Be^+, B^{2+}, and C^{3+}.

b) Why does δ_s decrease from Li to C^{3+}?

c) Excitation of Li from its ground state into the excited electronic configuration $1s^2 2p$ requires an energy of $1.848\ eV$. Determine the quantum defect δ_p.

d) Give a physical argument for the difference between δ_s and δ_p.

e) The experimental excitation energies for the $2s \rightarrow 3p$ and $2s \rightarrow 4p$ excitations of Li are $3.834\ eV$ and $4.522\ eV$, respectively. Use the Coulomb approximation to compute these excitation energies.

Solution :

a) The electron is removed from the $2s$ orbital and the ionization potentials, $IP(2s)$, are calculated from the expression

$$IP(2s) = Z_{net}^2 / 2(n^*)^2 \ , \ n^* = n - \delta_s$$

	$Li(Li\ I)$	$Be^+(Be\ II)$	$B^{2+}(B\ III)$	$C^{3+}(C\ IV)$
Z_{net}	1	2	3	4
n^*	1.5888	1.7289	1.7970	1.8375
δ_s	0.4112	0.2711	0.2030	0.1625

b) Large ion charge implies that the $1s$ orbitals become contracted; that is, the $1s$ electrons will almost screen with their whole charge. The potential seen by the optical $2s$ electron will thus be more and more coulomb–like when Z_{net} gets larger. Thus, δ_s becomes smaller when Z_{net} increases.

c) The excitation energy $\Delta E(2s \rightarrow 2p)$ is (for Li)

$$\Delta E(2s \rightarrow 2p) = -\frac{1}{2} \left[\frac{1}{(2 - \delta_p)^2} - \frac{1}{(2 - \delta_s)^2} \right] = 1.848\ eV$$

Since $\delta_s = 0.4112$ therefore $\delta_p = 0.0401$.

d) The reason for the decrease $\delta_s \rightarrow \delta_p$ may be interpreted as follows: The larger ℓ is, the larger is the probability for observing the optical electron at large distances from the nucleus and the more coulomb–like is the potential seen by the optical electron.

e)

$$\Delta E(2s \to 3p) = -\frac{1}{2}\left[\frac{1}{(3-\delta_p)^2} - \frac{1}{(2-\delta_s)^2}\right] = 3.837 \ eV$$

$$(\Delta E_{exp} = 3.834 \ eV)$$

$$\Delta E(2s \to 4p) = -\frac{1}{2}\left[\frac{1}{(4-\delta_p)^2} - \frac{1}{(2-\delta_s)^2}\right] = 4.522 \ eV$$

$$(\Delta E_{exp} = 4.522 \ eV)$$

Thus, the Coulomb approximation describes the experimental spectrum very well.

Example - 6.3 :

According to the first order perturbation treatment an electron in the He atom (call it electron 1) moves under the influence of a nuclear charge $Z = 2$ and the time average charge distribution of the other electron. The wave equation for electron 1 is: $[-\nabla_1^2 - 2Z_{eff}/r_1]u(1) = Eu(1)$, where $-2Z_{eff}/r_1 = -2Z/r_1 + V(r_1)$.
a) Obtain an expression for $V(r)$.
b) Plot Z_{eff} with $Z = 2$ in the range $r = 0$ to 2 and discuss the results.

Solution :

a) According to the perturbation theory the energy of repulsion of the two electrons in He is equal to $2I =< 1s(1)1s(2)|2/r_{12}|1s(1)1s(2) >$. This integral has the form:

$$2I = \int_0^\infty V(r_1)u_{1s}^2 4\pi r_1^2 dr_1 \quad ; \quad V(r_1) = \frac{2}{r_1}\left[1 - e^{-2Zr_1}(1 + Zr_1)\right]$$

This function gives the potential of electron 1 in the field of electron 2. The total potential is $-2Z/r_1 + V(r_1)$.
b) By definition

$$Z_{eff} = Z - \frac{r_1}{2}V(r_1) = 1 + e^{-4r_1}(1 + 2r_1)$$

For small values of r_1, electron 2 provides little shielding and electron 1 sees the full nuclear charge. However, for $r_1 > 1$ the shielding (or screening) is essentially complete and electron 1 experiences only one unit of charge.

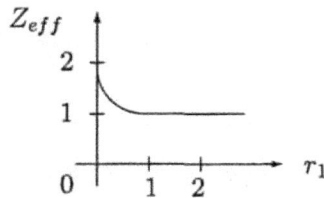

Example - 6.4 :

Derive an equation for the expectation value of the energy of the $1s^2 2s$ configuration of Li in terms of the integrals:
$J(1s, 2s) =< 1s(1)2s(2)|2/r_{12}|1s(1)2s(2) >$
$K(1s, 2s) =< 1s(1)2s(2)|2/r_{12}|2s(1)1s(2) >$.
In this derivation use a single Slater determinant to represent the state $1s^2 2s$. This determinant should be written out in full and the expectation value expression then multiplied out and simplified.

Solution :

$E =< H >=< \psi| \sum_{i=1}^{3} H_0(i) + H_1|\psi >$, $H_0(i) = -\nabla_i^2 - 6/r_i$, $H_1 = \sum_{i<j}^{3}(2/r_{ij})$ in $a.u.$ (Rydberg) and

$$\psi = \frac{1}{\sqrt{3!}} \begin{vmatrix} 1s\alpha(1) & 1s\alpha(2) & 1s\alpha(3) \\ 1s\beta(1) & 1s\beta(2) & 1s\beta(3) \\ 2s\alpha(1) & 2s\alpha(2) & 2s\alpha(3) \end{vmatrix}$$

$$= \frac{1}{\sqrt{6}} (1s\alpha(1)1s\beta(2)2s\alpha(3) + 2s\alpha(1)1s\alpha(2)1s\beta(3) + 1s\beta(1)2s\alpha(2)1s\alpha(3))$$

$$-\frac{1}{\sqrt{6}} (1s\alpha(1)2s\alpha(2)1s\beta(3) + 1s\beta(1)1s\alpha(2)2s\alpha(3) + 2s\alpha(1)1s\beta(2)1s\alpha(3))$$

Using the orthogonality conditions for the space and spin functions,

$$< u_i\chi_k(1)|u_j\chi_\ell(1) >=< u_i(1)|u_j(1) >< \chi_k(1)|\chi_\ell(1) >= \delta_{ij}\delta_{k\ell}$$

$$E = <H> = \frac{1}{6}\sum_{i=1}^{3}[4 < 1s(i)|H_0|1s(I)> + 2 < 2s(i)|H_0|2s(I)>]$$

or $E = <H> = 2I(1s) + I(2s)$.

In order to calculate two–electron integrals, we factor out the common spin functions:

$\psi = \psi_1\alpha(1)\beta(2)\alpha(3) + \psi_2\beta(1)\alpha(2)\alpha(3) + \psi_3\alpha(1)\alpha(2)\alpha(3) = \sum_i \psi_i S_i$;

where

$$\psi_1 = \frac{1}{\sqrt{6}}[1s(1)1s(2)2s(3) - 2s(1)1s(2)1s(3)],$$

$$\psi_2 = \frac{1}{\sqrt{6}}[1s(1)2s(2)1s(3) - 1s(1)1s(2)2s(3)],$$

$$\psi_3 = \frac{1}{\sqrt{6}}[2s(1)1s(2)1s(3) - 1s(1)2s(2)1s(3)].$$

$$< H_1 > = < \psi|H_1|\psi >$$

$$= < \psi_1 S_1 + \psi_2 S_2 + \psi_3 S_3| \sum_{i<j}^{3} \frac{2}{r_{ij}}|\psi_1 S_1 + \psi_2 S_2 + \psi_3 S_3 >$$

$$= < \psi_1|H_1|\psi_1 > + < \psi_2|H_1|\psi_2 > + < \psi_3|H_1|\psi_3 >$$

since $< S_i|S_j > = \delta_{ij}$. Evaluating $< \psi_1|H_1|\psi_1 >$ explicitly gives:

$6 < \psi_1|H_1|\psi_1 > = < 1s(1)1s(2)2s(3) - 2s(1)1s(2)1s(3)|2/r_{12} + 2/r_{13} + 2/r_{23}|1s(1)1s(2)2s(3) - 2s(1)1s(2)1s(3) >$

$= < 1s(1)1s(2)|2/r_{12}|1s(1)1s(2) > + < 2s(1)1s(2)|2/r_{12}|2s(1)1s(2) >$
$+ < 1s(1)2s(3)|2/r_{13}|1s(1)2s(3) > + < 2s(1)1s(3)|2/r_{13}|2s(1)1s(3) >$
$+ < 1s(1)2s(3)|2/r_{23}|1s(2)2s(3) > + < 1s(2)1s(3)|2/r_{23}|1s(2)1s(3) >$
$-2 < 1s(1)2s(3)|2/r_{13}|2s(1)1s(3) >$

The terms $< \psi_2|H_1|\psi_2 >$ and $< \psi_3|H_1|\psi_3 >$ give exactly the same results, and we conclude that:

$< H_1 > = J(1s, 1s) + 2J(1s, 2s) - K(1s, 2s)$. Therefore,

$E = <H> = 2I(1s) + I(2s) + J(1s, 1s) + 2J(1s, 2s) - K(1s, 2s)$.

Example - 6.5 :

The first order energy of a hydrogen atom in the magnetic field $\mathbf{B}$ is given by $E_1^{(1)} = (e^2 B)/(12 m_e) < 1s|r^2|1s >$. The molar diamagnetic susceptibility χ_m is defined by the equation $\chi_m = -(N_A)/(\mu_0 H)(\partial E_1^{(1)}/\partial H)$. Here $B = \mu_0 H$, $\mu_0 = 4\pi \times 10^{-7}$ N/A^2 in SI units. For many–electron atoms the susceptibility can be approximately calculated by assuming that the electrons move independently. Estimate the diamagnetic susceptibility of He by assuming that:
a) $\psi_{He} = 1s(1)1s(2)$ with $Z = 2$,
b) $\psi_{He} = 1s(1)1s(2)$ with Z determined so as to minimize the energy.

Solution :

a)

$$E_1^{(1)} = \frac{e^2 B^2}{12 m_e}[< 1s|r_1^2|1s > + < 1s|r_2^2|1s >] = \frac{e^2 B^2}{6 m_e} < 1s|r^2|1s >$$

$$\frac{\partial E_1^{(1)}}{\partial H} = \frac{\partial}{\partial H}\left[\frac{e^2}{6 m_e} < 1s|r^2|1s > \mu_0^2 H^2\right] = \frac{e^2 \mu_0^2 H}{3 m_e} < 1s|r^2|1s >$$

$$\chi_m = -N_A \frac{e^2 \mu_0}{3 m_e} < 1s|r^2|1s > = -\frac{N_A e^2 \mu_0 a_0^2}{4 m_e}$$

Here we used $< 1s|r^2|1s > = 3a_0^2/Z^2$, $Z = 2$. Considering $\mu_0 \varepsilon_0 = 1/c^2$ we may write

$$\chi_m = -4\pi \left[\frac{N_A e^2 a_0^2}{(4\pi\varepsilon_0) 4 m_e c^2}\right]$$

Therefore, $\chi_m = -1.49 \times 10^{-11}$ m^3 .

b) From energy minimization, $Z_{eff} = 1.6875 \cong 1.688$

$$\psi = \left(\frac{Z_e^3}{\pi}\right)^{1/2} e^{-Z_e(r_1 + r_2)} \quad \rightarrow \quad E = Z_e^2 - 2Z_e Z + \frac{5}{8} Z_e$$

$$\frac{\partial E}{\partial Z_e} = 2Z_e - 2Z + \frac{5}{8} \quad \rightarrow \quad Z_e = Z - \frac{5}{16} = \frac{27}{16} = 1.6875 \cong 1.688$$

Thus the corresponding diamagnetic susceptibility is

$$\chi_m = -\frac{N_A e^2 \mu_0 a_0^2}{m_e (1.688)^2} = -2.09 \times 10^{-11} \; m^3$$

The experimental value is $-2.36 \times 10^{-11} \; m^3$.

Example - 6.6 :

Derive the Russell–Saunders terms for 9 equivalent d–electrons, i.e. for the configuration d^9.

Solution :

The possible combinations of m_l and m_s for d electrons are represented by the boxes below. The number of ways that 9 equivalent electrons can be placed in these boxes is equal to the number of locations of the "hole". Therefore, ten states are predicted.

$m_l =$	+2	+1	0	−1	−2
$m_s = \quad +1/2$					
$-1/2$					

The maximum possible value for $\sum_{i=1}^{9} m_{l_i}$ is 2 which occurs when the missing electron is assigned to $m_l = -2$. It is clear that the $M_L = +2$ state has $M_S = 1/2$ since there is only one unpaired electron. We conclude that the only term is ^2D. The d^9 configuration is an example of a general rule. The terms which arise for the configuration x^r are the same as those that arise for $x^{2(2l+1)-r}$, e.g. d^1 and d^9 give the same terms.

Example - 6.7 :

List the lowest terms for the following atoms:
a) Be , b) C , c) O , d) Cl , e) As

Solution :

a) $Be[1s^2 2s^2]$: Since the shells are closed, the term is $^1S_0$.

b) $C[1s^2 2s^2 2p^2]$: The states of maximum multiplicity have $S = 1$ and of these the maximum L is 1. Therefore, the lowest term is $^3P_0$. Since the shell is less than half filled, the lowest energy corresponds to the lowest value of J, i.e. the multiplet is regular.

c) $O[1s^2 2s^2 2p^4]$: The maximum value of S is 1 and the largest L consistent with this spin state is 1. Therefore, the lowest term is $^3P_2$. Here the lowest term has the maximum value of J since the shell is more than half filled, i.e. the multiplet is inverted.

d) $Cl[1s^2 2s^2 2p^6 3s^2 3p^5]$: Hund's rules for this configuration give $S = 1/2$ and $L = 1$. The highest value of J is 3/2. Therefore the lowest term is $^3P_{3/2}$.

e) $As[1s^2 2s^2 2p^6 3s^2 3p^6 4s^2 4p^3]$: The maximum S is 3/2 and $L = 0$. Therefore, the lowest term is $^4S_{3/2}$.

Example - 6.8 :

Calculate approximately the frequency and wavelength of K_α–line of molybdenum, and the energy of a quantum corresponding to this line.

Solution :

$\nu = Rc(Z - 1)^2 \left(\frac{1}{1} - \frac{1}{2^2} \right) = 4.16 \times 10^{18} \ s^{-1}$

$\lambda_\alpha = 0.72 \ \mathring{A} \ , \ E = h\nu = 17.10 \ eV$.

Chapter 7

Molecular Structure

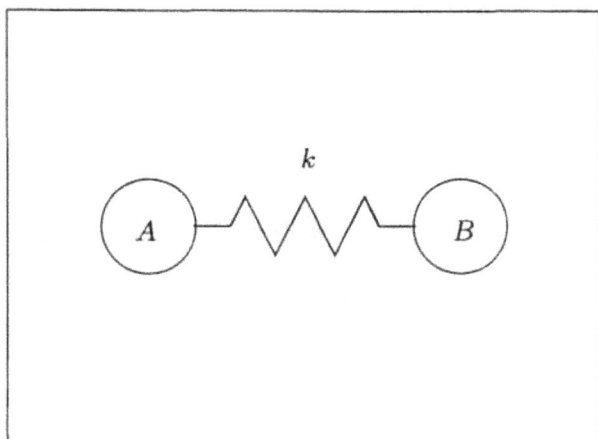

7.1   The Schrödinger equation for molecules

Molecular wave functions Ψ_E and energies E are obtained by solving the appropriate time–independent Schrödinger equation

$$H\Psi_E = E\Psi_E \tag{7.1}$$

This equation is more complicated for a molecular than an atomic system. The molecular Hamiltonian contains more interaction terms. In general, the non–relativistic Hamiltonian is simply the sum of the kinetic and potential energy operators,

$$H = T + V \tag{7.2}$$

We write the kinetic energy operator for a molecule consisting of N electrons and M nuclei as

$$T = -\frac{\hbar^2}{2m}\sum_{i=1}^{N}\nabla_i^2 - \frac{\hbar^2}{2}\sum_{k=1}^{M}\frac{1}{M_k}\nabla_k^2 = T_e + T_n \tag{7.3}$$

here T_e and T_n are the kinetic energy of electrons and nuclei, respectively.

The potential energy consists of several interactions: Coulombic interactions, spin–orbit, and other magnetic interactions, and so on. From our study of atoms, we expect the contributions of the magnetic terms to the system's energy to be small, since they are relativistic corrections. So for the moment we will drop all terms from the potential energy except those due to the Coulomb interaction. The Coulomb terms are three types: V_{ee}, the potential energy due to interaction between electrons, V_{en}, the potential energy due to interactions between electrons and nuclei, and V_{nn}, the potential energy due to interaction between nuclei. Symbolically, we have

$$V = V_{ee} + V_{en} + V_{nn} \tag{7.4}$$

We write the explicit form of these terms again for a molecule consisting of N electrons and M nuclei as

$$V_{ee} = \sum_{i<j}^{N}\frac{e^2}{r_{ij}} \tag{7.5}$$

$$V_{en} = -\sum_{i=1}^{N}\sum_{k=1}^{M}\frac{Z_k e^2}{r_{ik}} \qquad (7.6)$$

$$V_{nn} = \sum_{k<l}^{N}\frac{Z_k Z_l e^2}{r_{kl}} \qquad (7.7)$$

The coordinates of the electrons in a molecule are usually represented by the set $\{\mathbf{r}_i\}$, $i = 1, 2, ..., N$, on the other hand the coordinates of the nuclei are usually represented by the set $\{\mathbf{R}_k\}$, $i = 1, 2, ..., M$. Therefore the coordinate dependence of each term in the molecular Hamiltonian may be indicated explicitly as

$$H(\mathbf{r}_i, \mathbf{R}_k) = T_e(\mathbf{r}_i) + T_n(\mathbf{R}_k) + V_{ee}(\mathbf{r}_i) + V_{en}(\mathbf{r}_i, \mathbf{R}_k) + V_{nn}(\mathbf{R}_k) \quad (7.8)$$

This Hamiltonian describes the quantum mechanical system, the molecule. The stationary states of this system are described by the time–independent molecular wave function Ψ_E, which contain information about the dynamics of all particles forming the molecule. Since H depends on electronic coordinates $\mathbf{r}_i$ and nuclear coordinates $\mathbf{R}_k$, we expect that Ψ_E also depends on both sets of coordinates; that is

$$\Psi_E = \Psi_E(\mathbf{r}_i, \mathbf{R}_k) \qquad (7.9)$$

These eigenfunctions and corresponding energies are obtained by solving the Schrödinger equation, Eq.(7.1).

7.2 The Born–Oppenheimer approximation

In a classical description the electrons move very fast around the heavy nuclei. In the time it takes a nucleus to move a short distance, we expect the electrons to be able to traverse the entire molecule, perhaps several times. Such a picture suggests that at any particular time the electrons' dynamics depend primarily on the location of the nuclei and not so much on their velocities. Since the time required for the electrons to traverse their orbits is so much shorter than the time characteristic of nuclear motion, therefore from the point of view of the electrons, the nuclei can be regarded as fixed in space.

This picture is the basis of the Born–Oppenheimer approximation. The Born–Oppenheimer approximation was first formulated by Max Born and J. Robert Oppenheimer in 1927. In this approximation, we first obtain functions that describe the motion of the electrons keeping the nuclei fixed in space. We then consider the motion of the nuclei themselves and obtain functions describing that motion.

The Born–Oppenheimer approximation may be formulated briefly as follows:

Considering the nuclei as infinitely massive and hence fixed in space at positions $\mathbf{R}_k$, we study the electronic motion. In this consideration $T_n(\mathbf{R}_k) = 0$. The only terms in the molecular Hamiltonian that are not constant at this point are T_e, V_{ee}, and V_{en}. They are used to define the so–called electronic Hamiltonian.

$$H_e(\mathbf{r}_i, \mathbf{R}_k) \equiv T_e(\mathbf{r}_i) + V_{ee}(\mathbf{r}_i) + V_{en}(\mathbf{r}_i, \mathbf{R}_k) \tag{7.10}$$

We now solve the Schrödinger equation for this Hamiltonian, the so–called electronic Schrödinger equation,

$$H_e \Psi^e_{E_e}(\mathbf{r}_i, \mathbf{R}_k) = E_e(\mathbf{R}_k) \Psi^e_{E_e}(\mathbf{r}_i, \mathbf{R}_k) \tag{7.11}$$

for the electronic eigenfunctions $\Psi^e_{E_e}(\mathbf{r}_i, \mathbf{R}_k)$ and electronic energies $E_e(\mathbf{R}_k)$. The constant nuclear coordinates appear explicitly in the electronic wave functions as parameters. Because of the presence of the electron–nucleus interaction term $V_{en}(\mathbf{r}_i, \mathbf{R}_k)$ in the electronic Hamiltonian, the electronic energy $E_e(\mathbf{R}_k)$ is a function of all the nuclear coordinates $\mathbf{R}_k$, which were all kept constant while solving for the electronic motion. This is an example of an adiabatic approximation. In fact, the Born–Oppenheimer approximation is sometimes called the **adiabatic approximation**.

Separation of molecular energy levels:

After obtaining electronic states we combine the electronic energy with the Coulomb nuclear repulsion terms to form a total nuclear potential energy. This term is used to define a nuclear Hamiltonian

$$H_n(\mathbf{R}_k) \equiv T_n(\mathbf{R}_k) + V_{nn}(\mathbf{R}_k) + E_e(\mathbf{R}_k) \tag{7.12}$$

The time–independent nuclear Schrödinger equation

$$H_n(\mathbf{R}_k)\Psi_{E_n}(\mathbf{R}_k) = E_n\Psi_{E_n}(\mathbf{R}_k) \tag{7.13}$$

is then solved for the nuclear wave function $\Psi_{E_n}(\mathbf{R}_k)$ and energies E_n. The energy E_n is a constant, but the nuclear wave function is a function of the nuclear coordinates $\mathbf{R}_k$.

The Born–Oppenheimer approximation states that the stationary-state wave functions of the molecule can be approximated by products of the solutions to the electronic and nuclear motions. In particular, a good approximate wave function for the molecule is given by

$$\Psi_E(\mathbf{r}_i, \mathbf{R}_k) = \psi_{E_e}(\mathbf{r}_i, \mathbf{R}_k)\psi_{E_n}(\mathbf{R}_k) \tag{7.14}$$

7.3 Rotational and vibrational motions of diatomic molecules

For a diatomic molecule $A-B$ with n electrons, the complete Schrödinger equation which includes both nuclear and electronic motion is

$$\left[-\frac{\hbar^2}{2\mu}\nabla^2 - \frac{\hbar^2}{2m}\sum_{i=1}^{n}\nabla_i^2 + V \right]\Psi = E\Psi \tag{7.15}$$

where Ψ is the total electronic and nuclear wave function, and E is the total energy; and the reduced mass μ is defined as

$$\frac{1}{\mu} = \frac{1}{M_A} + \frac{1}{M_B} \tag{7.16}$$

$$V = V_{ee} + V_{eN} + V_{NN} \tag{7.17}$$

Electron–electron repulsion energy:

$$V_{ee} = \sum_{i<j}^{n} \frac{e^2}{r_{ij}} \tag{7.18}$$

Electron–nuclear attraction energy:

$$V_{eN} = -\sum_{i=1}^{n} \frac{Z_A e^2}{r_{i_A}} - \sum_{i=1}^{n} \frac{Z_B e^2}{r_{i_B}} \tag{7.19}$$

Nuclear–nuclear repulsion energy:

$$V_{NN} = \frac{Z_A Z_B e^2}{R} \tag{7.20}$$

$$\Psi = \Psi(\mathbf{r}, \mathbf{R}) \quad , \quad \mathbf{r} \equiv (\mathbf{r}_1, \mathbf{r}_2, ..., \mathbf{r}_n) \tag{7.21}$$

Born–Oppenheimer approximation:

$$\Psi(\mathbf{r}, \mathbf{R}) = \psi(\mathbf{r}, \mathbf{R})\chi(\mathbf{R}) \tag{7.22}$$

$\psi(\mathbf{r}, \mathbf{R})$ and $\chi(\mathbf{R})$ are the electronic and nuclear wave functions, respectively.

Electronic Schrödinger equation:

$$\left[-\frac{\hbar^2}{2m} \sum_{i=1} n\nabla_i^2 + V(\mathbf{r}, \mathbf{R}) \right] \psi(\mathbf{r}, \mathbf{R}) = E(R)\psi(\mathbf{r}, \mathbf{R}) \tag{7.23}$$

Therefore

$$\left[-\frac{\hbar^2}{2\mu}\nabla^2 + E(R) \right] \psi(\mathbf{r}, \mathbf{R})\chi(\mathbf{R}) = E\psi(\mathbf{r}, \mathbf{R})\chi(\mathbf{R}) \tag{7.24}$$

or

$$\left[-\frac{\hbar^2}{2\mu}\nabla^2 + E(R) \right] \chi(\mathbf{R}) = E\chi(\mathbf{R}) \tag{7.25}$$

is the Schrödinger equation for the nuclear motion of a diatomic molecule.

In spherical polar coordinates

$$\chi(R, \theta, \phi) = \mathcal{R}(R)\mathcal{S}(\theta, \phi) \equiv \mathcal{R}(R)\mathcal{S}_{JM}(\theta, \phi) \tag{7.26}$$

$\mathcal{R}(R)$ and $\mathcal{S}_{JM}(\theta, \phi)$ are the radial and angular wave functions, respectively. J is the molecular total angular momentum quantum number and M is its z–component. Let

$$\mathcal{H} = -\frac{\hbar^2}{2\mu R^2} \left[\frac{\partial}{\partial R} \left(R^2 \frac{\partial}{\partial R} \right) + \frac{1}{\sin\theta} \frac{\partial}{\partial\theta} \left(\sin\theta \frac{\partial}{\partial\theta} \right) + \frac{1}{\sin^2\theta} \frac{\partial^2}{\partial\phi^2} \right]$$

$$[\mathcal{H} + E(\mathbf{R})]\chi(R, \theta, \phi) = E\chi(R, \theta, \phi) \tag{7.27}$$

The functions $S_{JM}(\theta, \phi)$ are the eigenfunctions of the operator

$$M^2 = -\hbar^2 \left(\frac{1}{\sin\theta} \frac{\partial}{\partial\theta} \sin\theta \frac{\partial}{\partial\theta} + \frac{1}{\sin^2\theta} \frac{\partial^2}{\partial\phi^2} \right) \tag{7.28}$$

$$M^2 S_{JM}(\theta, \phi) = \hbar^2 J(J+1) S_{JM}(\theta, \phi) \tag{7.29}$$

Therefore

$$\left\{ -\frac{\hbar^2}{2\mu R^2} \left[\frac{\partial}{\partial R} R^2 \frac{\partial}{\partial R} - J(J+1) \right] + E(R) \right\}$$

$$\times \mathcal{R}(R) S_{JM}(\theta, \phi) = E_{v,J} \mathcal{R}(R) S_{JM}(\theta, \phi) \tag{7.30}$$

here v is the vibrational quantum number, and J is the rotational quantum number.

Rewriting the Schrödinger equation

$$\left[-\frac{\hbar^2}{2\mu R^2} \frac{d}{dR} \left(R^2 \frac{d}{dR} \right) + \frac{J(J+1)\hbar^2}{2\mu R^2} + E(R) \right] \mathcal{R}(R) = E_{v,J} \mathcal{R}(R) \tag{7.31}$$

which is **radial Schrödinger equation** for the nuclear motion. Its solutions $\mathcal{R}(R)$ give the vibrational wave functions, while $S_{JM}(\theta, \phi)$ give the rotational wave functions.

Rotational energy levels:

In rigid rotator approximation:

$$E_{rot} = E_J = \frac{\hbar^2}{2I_e} J(J+1) = \frac{\hbar^2}{2\mu R_e^2} J(J+1) = BJ(J+1) \ , \ \ J = 0, 1, 2, ... \tag{7.32}$$

In non–rigid rotator approximation:

$$E_{rot} = BJ(J+1) - D[J(J+1)]^2 \tag{7.33}$$

here B and D are called rotational and centrifugal distortion constants, respectively.

Vibrational energy levels:

In harmonic approximation:

$$E_{vib} = E_v = (v + \frac{1}{2})h\nu_e \quad , \quad v = 0, 1, 2, \ldots \tag{7.34}$$

In anharmonic approximation:

$$E_v = (v + \frac{1}{2})h\nu_e - (v + \frac{1}{2})^2 h\nu_e x_e \tag{7.35}$$

here k_e and x_e are called force and anharmonicity constants, respectively. The fundamental frequency is defined as $\nu_e = (1/2\pi)\sqrt{k_e/\mu}$.

Energy of nuclear motion:

$$E_{nuc} = E_{v,J} = E_{vib} + E_{rot} \tag{7.36}$$

In general,

$$E_{nuc} = (v + \frac{1}{2})h\nu - (v + \frac{1}{2})^2 x_e h\nu + \frac{\hbar^2}{2I}J(J+1) \tag{7.37}$$

$$-\frac{\hbar^6}{2I^3(h\nu)^2}J^2(J+1)^2 - \alpha(v + \frac{1}{2})J(J+1)$$

here α is called the vibration–rotation coupling constant. $[D] \sim [B] \times 10^{-3}$ and $[\alpha] \sim [B] \times 10^{-2}$. In practice, experimentally the energy of nuclear motion is measured in units of cm^{-1}. Therefore the wave number $\tilde{\nu}$ corresponding to any nuclear motion energy $E_{v,J}$ is expressed in cm^{-1} as

$$\tilde{\nu} = \frac{E_{v,J}}{hc} = \tilde{\nu}(v + \frac{1}{2}) - \tilde{\nu}x_e(v + \frac{1}{2})^2 + \tilde{B}J(J+1) \tag{7.38}$$

$$-\tilde{D}J^2(J+1)^2 - \alpha(v + \frac{1}{2})J(J+1) \quad (in \; cm^{-1})$$

here

$$\tilde{\nu} = \frac{\nu}{c} \quad , \quad \tilde{B} = \frac{\hbar^2}{2hcI} \quad , \quad \tilde{D} = \frac{4\tilde{B}^3}{\tilde{\nu}^2}$$

Total energy of a molecule:

$$E_r = E_e + E_v + E_r \tag{7.39}$$

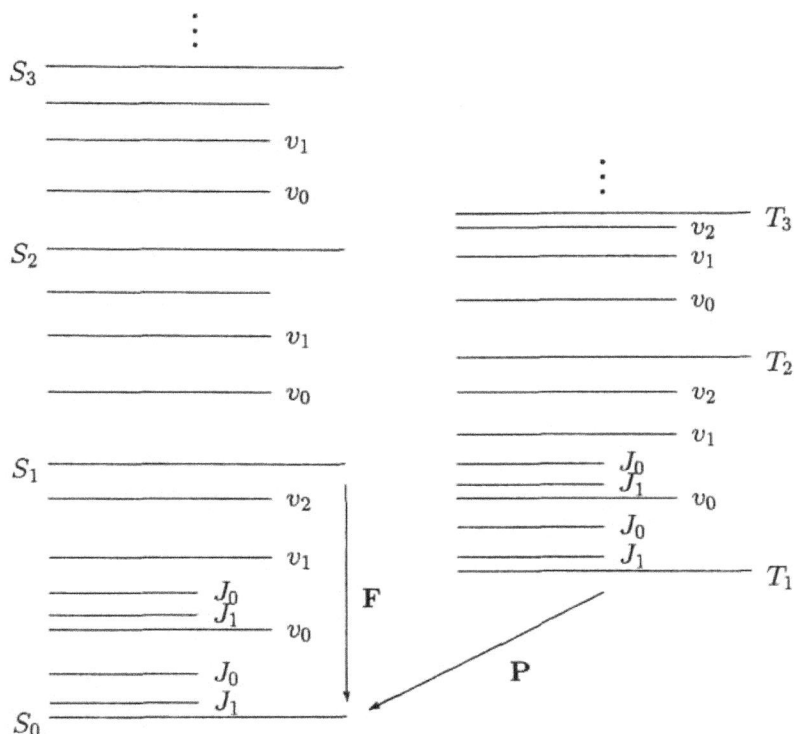

F: Fluorescence (spin allowed)
P: Phosphorescence (spin forbidden)
Fig. 7.1. Relative positions of the energy levels of a molecule.

7.4 Atomic interactions

Since atoms are composed of charged particles, we expect the inter-atomic forces to be largely electrostatic in nature. The interaction of two atoms will depend upon (1) the charge state of the atom (whether they are neutral or ions) and (2) the electronic structure of the atoms (or ions).

Atomic interactions may be classified whether their outermost electron shell is complete (closed–shell atoms) or incomplete (open–shell atoms).

For the discussion of the molecular bonding, it is convenient to make the following classification of interactions:

1. Interaction between a closed–shell positive ion and a closed–shell negative ion; for example, Na^+ and Cl^-. This gives rise to an **ionic bond**. A^+(closed–shell)B^-(closed–shell).

2. Interaction between two closed–shell neutral atoms; for example, $He - He$. This is the **Van der Waals interaction**. A(closed–shell)B(closed–shell).

3. Interaction between an open–shell ion (commonly positive) and one or more closed–shell ions (commonly negative). This occurs, for example, in transition–metal complexes of the type $M^{2+}(X^-)_6$. A^{n+}(open–shell)$(B^-)_m$(closed–shell).

4. Interactions between two open–shell neutral atoms; for example, $H - H$. This gives rise to the **covalent bond**. A(open–shell)B(open–shell).

Although this classification is general, there are some exeptions; for example, a bond between H and F involves two open–shell atoms (type 4) but there are significant ionic contributions to the interaction energy (type 1).

7.5   The ionic bond

The ionization potentials of the alkali metals are small, while the electron affinities of the halogens are large. Thus, the positive ions Li^+, Na^+, K^+, Rb^+, Cs^+, and the negative ions F^-, Cl^-, Br^-, I^- are relatively easy to make. When brought together, such positive and negative ions interact strongly and form a bond that is essentially ionic.

Consider Na^+Cl^-:

$$Na \;\rightarrow\; Na^+ + e^- \;\; : \;\; IP_{Na} > 0 \tag{7.40}$$

$$e^- + Cl \;\rightarrow\; Cl^- \;\; : \;\; EA_{Cl} < 0 \tag{7.41}$$

$$Na+Cl \rightarrow Na^+ + Cl^- \quad : \quad \Delta E + IP_{Na} + EA_{Cl} = \Delta E(\infty) > 0 \quad (7.42)$$

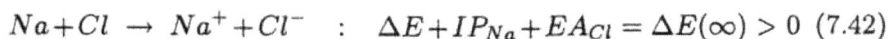

$$or \ \ \Delta E = IP_{Na} - |EA_{Cl}| = \Delta E_\infty$$

As the ions are brought together, the ion–pair system is stabilized by the electrostatic interaction energy. At an interionic separation R, sufficiently large to neglect overlap effect, we have

$$\Delta E(R) = -\frac{e^2}{R} + \Delta E(\infty) \qquad (7.43)$$

At small interionic distances, the electron clouds of the two ions overlap. Because the ions have closed–shell configurations, the interaction between them is repulsive, we can approximate it by a simple exponential of the form $Ae^{-\alpha R}$ (Born–Mayer potential). Therefore we write the energy of the ion pair relative to that of the neutral atoms as

$$\Delta E(R) = Ae^{-\alpha R} - \frac{e^2}{R} + \Delta E(\infty) \qquad (7.44)$$

This is a semi empirical potential. Here the constanta A and α can be determined theoretically from the wave functions for the atoms or estimated experimentally by fitting known properties of the $NaCl$ molecule.

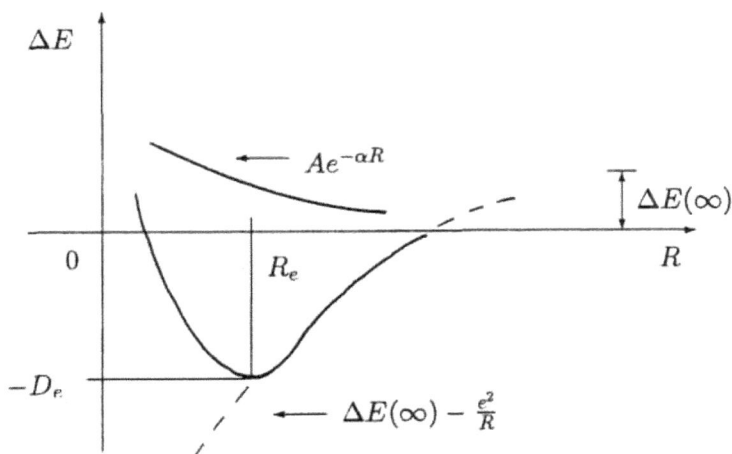

Fig. 7.2. Semi–empirical potential.

R_e is the equilibrium internuclear distance, D_e is the bond energy (or dissociation energy).

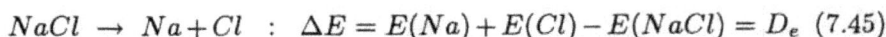

$$NaCl \rightarrow Na + Cl \quad : \quad \Delta E = E(Na) + E(Cl) - E(NaCl) = D_e \quad (7.45)$$

$$NaCl \rightarrow Na^+ + Cl^- \quad : \quad \Delta E = E(Na^+) + E(Cl^-) - E(NaCl) \quad (7.46)$$

$$\Delta E = D_e + IP - EA$$

Another property of the potential energy function (PEF) that can be determined experimentally is the value of the second derivative of the interaction energy $\Delta E(R)$ at $R = R_e$ which is equal to the force constant k_e in the harmonic oscillator approximation.

For a harmonic oscillator the vibration frequency ν_e is related to k_e by the equation

$$\nu_e = \frac{1}{2\pi} \sqrt{\frac{k_e}{\mu}} \quad (7.47)$$

where μ is the reduced mass associated with the relative motion of the two atoms. The vibrational frequency is usually expressed in wave numbers (cm^{-1}) and given the symbol ω_e.

We now have three experimental numbers; R_e, D_e, and k_e.

$$\Delta E(R_e) = -D_e \quad (7.48)$$

$$\frac{\partial \Delta E(R)}{\partial R}\Big|_{R=R_e} = 0 \quad (7.49)$$

$$\frac{\partial^2 \Delta E(R)}{\partial R^2}\Big|_{R=R_e} = k_e \quad (7.50)$$

Considering these equations the unknown parameters in the PEF can easily be determined.

7.6   The van der Waals attraction

Let the two *one-dimensional* one-electron atoms approach to an internuclear distance R which is close enough for the electrons to interact with each other and the opposite nuclei, but not so close as to cause any appreciable repulsion due to the overlapping of the electron clouds.

Fig. 7.3. One–dimentional molecular model.

The interaction potential V' between the two atoms is the sum of the electrostatic potentials for the pairs of particles considered.

$$V'(z_1, z_2) = e^2 \left(\frac{1}{R} + \frac{1}{R + z_2 - z_2} - \frac{1}{R - z_1} - \frac{1}{R + z_2} \right) \qquad (7.51)$$

expending V' about $z_1 = z_2 = 0$ gives

$$V'(z_1, z_2) = \qquad (7.52)$$

$$e^2 \left\{ \frac{1}{R} + \left[R \left(1 + \frac{z_2 - z_1}{R} \right) \right]^{-1} - \left[R \left(1 - \frac{z_1}{R} \right) \right]^{-1} - \left[R \left(1 + \frac{z_2}{R} \right) \right]^{-1} \right\}$$

$$= e^2 \left\{ \frac{1}{R} + \frac{1}{R} \left[1 - \left(\frac{z_2 - z_1}{R} \right) + \left(\frac{z_2 - z_1}{R} \right)^2 + \ldots \right] \right\}$$

$$- e^2 \left\{ \frac{1}{R} \left[1 + \frac{z_1}{R} + \left(\frac{z_1}{R} \right)^2 + \ldots \right] \right\}$$

$$- e^2 \left\{ \frac{1}{R} \left[1 - \frac{z_2}{R} - \left(\frac{z_2}{R} \right)^2 + \ldots \right] \right\}$$

$$= -\frac{2e^2 z_1 z_2}{R^3} + higher\ order\ terms\ (can\ be\ neglected) \qquad (7.53)$$

The total quantum mechanical Hamiltonian for the one–dimensional motion along the internuclear axis is

$$H = -\frac{\hbar^2}{2m} \frac{\partial^2}{\partial z_1^2} + \frac{1}{2} k z_1^2 - \frac{\hbar^2}{2m} \frac{\partial^2}{\partial z_2^2} + \frac{1}{2} k z_2^2 + V'(z_1, z_2) \qquad (7.54)$$

$$= -\frac{\hbar^2}{2m} \left(\frac{\partial^2}{\partial z_1^2} + \frac{\partial^2}{\partial z_2^2} \right) + \frac{1}{2} k(z_1^2 + z_2^2) - \frac{2e^2}{R^3} z_1 z_2$$

Hamiltonian of a two–dimensional harmonic oscillator (coupled oscillators) which is non–seperable.

Introducing the variables $\eta = z_2 - z_1$ and $\xi = z_1 + z_2$

$$H = -\frac{\hbar^2}{2m}\frac{\partial^2}{\partial \eta^2} + \frac{1}{4}\left(k + \frac{2e^2}{R^3}\right)\eta^2 - \frac{\hbar^2}{2m}\frac{\partial^2}{\partial \xi^2} + \frac{1}{4}\left(k - \frac{2e^2}{R^3}\right)\xi^2 \quad (7.55)$$

This Hamiltonian is separable. The effective force constants of the oscillators are

$$k_1 = \frac{1}{2}k\left(1 + \frac{2e^2}{kR^3}\right) \quad , \quad k_2 = \frac{1}{2}k\left(1 - \frac{2e^2}{kR^3}\right) \quad (7.56)$$

since $\nu = (1/2\pi)\sqrt{k/\mu}$

$$\nu_1 = \frac{1}{2\pi}\sqrt{\frac{k_1}{\mu}} = \frac{1}{2\pi}\sqrt{\frac{k_1}{m}\left(1 + \frac{2e^2}{kR^3}\right)} \quad (7.57)$$

$$\nu_2 = \frac{1}{2\pi}\sqrt{\frac{k_2}{\mu}} = \frac{1}{2\pi}\sqrt{\frac{k_1}{m}\left(1 - \frac{2e^2}{kR^3}\right)}$$

Expanding ν_1 and ν_2 about $(1/R) = 0$, gives

$$\nu_1 = \nu\left[1 + \frac{e^2}{kR^3} - \frac{1}{8}\left(\frac{2e^2}{kR^3}\right)^2 + \cdots\right] \quad (7.58)$$

$$\nu_2 = \nu\left[1 - \frac{e^2}{kR^3} - \frac{1}{8}\left(\frac{2e^2}{kR^3}\right)^2 + \cdots\right]$$

where ν is the frequency of the non–interacting oscillators.

The ground state of the oscillators is the zero–point energy

$$E_0 = \frac{1}{2}h(\nu_1 + \nu_2) = h\nu - \frac{1}{2}h\nu\frac{e^4}{k^2R^6} + \cdots \quad (7.59)$$

The second term is called as the **dispersion energy** which is attractive

$$\Delta E(R)_{dis} = -\frac{1}{2}h\nu\frac{e^4}{k^2R^6} \quad (7.60)$$

Considering the **polarizability** of a Hooke's law atom, $\alpha = e^2/k$,

$$\Delta E(R)_{dis} = -\frac{1}{2}h\nu\frac{\alpha^2}{R^6} \tag{7.61}$$

This energy may be interpreted as the interaction of two **fluctuating dipoles**.

In the simple treatment outlined above, we have restricted the oscillators to one–dimensional motion. If this restriction is removed, the result becomes

$$\Delta E(R)_{dis} = -\frac{3}{4}h\nu\frac{\alpha^2}{R^6} \tag{7.62}$$

A more realistic approximate treatment of the dispersion forces between two atoms gives the same result, Eq.(7.62), except that the oscillator energy $h\nu$ is replaced by the first IP of the atom, that is

$$\Delta E(R)_{dis} = -\frac{3}{4}(IP)\frac{\alpha^2}{R^6} \tag{7.63}$$

For a pair of unlike atoms A and B, the formula becomes

$$\Delta E(R)_{dis} = -\frac{3}{2}\left(\frac{IP_A \cdot IP_B}{IP_A + IP_B}\right)\frac{\alpha_A\alpha_B}{R^6} \tag{7.64}$$

This approximate equation was first derived by W. Heitler and F. London (in 1927); it is often written in the form

$$\Delta E(R)_{dis} = -\frac{C_6}{R^6} \tag{7.65}$$

where C_6 is called the Van der Waals coefficient.

The complete curve for a Van der Waals interaction $(He - He)$ may be written as

$$\Delta E(R) = Ae^{-\alpha R} - \frac{C_6}{R^6} \tag{7.66}$$

$$C_6 \cong \frac{3}{4}(IP)\alpha^2 \tag{7.67}$$

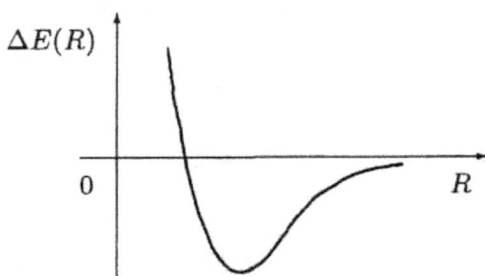

Fig. 7.4. Interaction energy vs interatomic separation.

Since the interaction energy for $He - He$ consists of both repulsive and attractive terms, the question of formation of a stable molecule He_2 arises. The internal vibrational energy (even at absolute zero) is larger than the binding energy and no stable molecule can exist. However, He_2 is stable relative to dissociation in excited electronic states, e.g., a molecule formed from $He(1s^2)$ and $He(1s2s)$, the spectrum of such molecules have been observed.

For closed–shell neutral atoms, the van der Waals force can be expected to be the principal attractive force. For open–shell atoms and ions there can be other attractive forces which are much stronger than those due to the Van der Waals interaction. However, the Van der Waals force is always present, whether the atoms are bonded in a molecule, in the same molecule but not directly bonded, or in different molecules.

Relative positions of vibrational levels for the ground electronic state in a pair potential:

D_e : Spectroscopic dissociation energy (calculated from vibration–rotation spectra)
D_0 : Chemical dissociation energy (calculated from thermal data)
$(r_0 > r_e)$

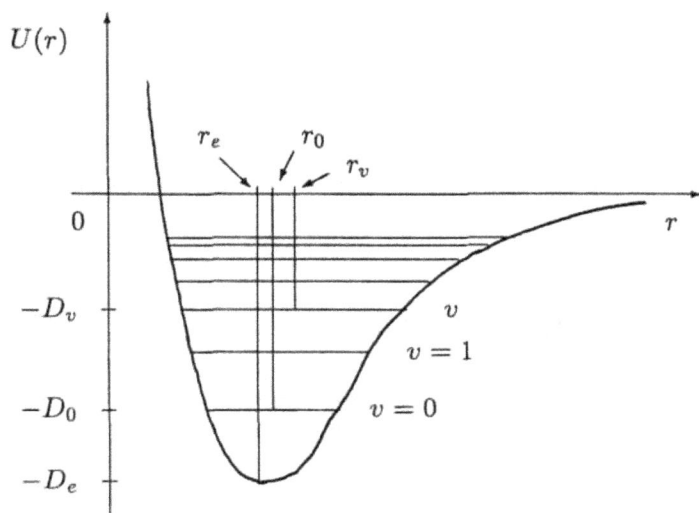

Fig. 7.5. Relative positions of vibrational levels.

Some pair potentials (Empirical interatomic potentials):

Born–Mayer potential (for ionic pairs)

$$U(r) = Ae^{-\alpha r} - \frac{B}{r} \tag{7.68}$$

Buckingham potential (for van der Waals interactions)

$$U(r) = Ae^{-\alpha r} - \frac{C_6}{r^6} \tag{7.69}$$

Mie potential

$$U(r) = \epsilon \left[\frac{n}{m-n} \left(\frac{r_0}{r} \right)^m - \frac{m}{m-n} \left(\frac{r_0}{r} \right)^n \right] \quad , \quad (m > n) \tag{7.70}$$

Lennard–Jones potential ($m = 12$, $n = 6$ in Mie potential)

$$U(r) = \epsilon \left[\left(\frac{r_0}{r} \right)^{12} - 2 \left(\frac{r_0}{r} \right)^6 \right] \tag{7.71}$$

Morse potential

$$U(r) = D_e \left[1 - e^{-\beta(r-r_0)}\right]^2 \tag{7.72}$$

Erkoç potential

$$U(r) = A\left[\left(\frac{r_0}{r}\right)^{2n} e^{-2\alpha(\frac{r}{r_0})^2} - \left(\frac{r_0}{r}\right)^n e^{-\alpha(\frac{r}{r_0})^2}\right] \tag{7.73}$$

7.7   Open–shell interactions

The interaction of two atoms, both of which have unfilled valence shell, leads to the most common type of molecular bond. Consider two hydrogen atoms in their ground states. As the two nuclei approach one another, their electron clouds overlap. If the electron spins are paired, the electron of each of the atoms can occupy the $1s$ orbital of the other, without voilating the Pauli principle, and thereby be near both nuclei.

The discussion of open–shell atoms leads naturally to the idea of **sharing** or **exchanging** of electrons.

We write down the potential energy function for two hydrogen atoms, a system consisting of two protons a distance R apart and two electrons. The total electrostatic potential energy function is

$$V = -\frac{e^2}{r_{1_A}} - \frac{e^2}{r_{2_B}} - \frac{e^2}{r_{2_A}} - \frac{e^2}{r_{1_B}} + \frac{e^2}{r_{12}} + \frac{e^2}{R} \tag{7.74}$$

The complete Hamiltonian for the system includes, in addition to V of Eq.(7.74), the kinetic energy terms for each of the four particles.

Since we are interested in the electronic properties of the system, we neglect the nuclear motion by assuming the nuclei to be fixed at the internuclear distance R. This is a valid approximation (Born–Oppenheimer approximation) and the motion of the nuclei can be treated after the electronic problem for fixed nuclei has been solved.

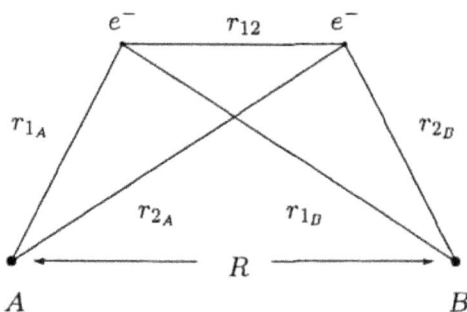

Fig. 7.6. Relative positions of nuclei and electrons in H_2.

The fixed–nucleus Hamiltonian is

$$H = -\frac{\hbar^2}{2m}(\nabla_1^2 + \nabla_2^2) + V \qquad (7.75)$$

Since the Schrödinger equation corresponding to the Hamiltonian cannot be solved exactly, we proceed by examining a number of different approximations.

At very large values of R we write the Hamiltonian for the two (non–interacting) atoms as

$$H_0 = -\frac{\hbar^2}{2m}\nabla_1^2 - \frac{e^2}{r_{1_A}} - \frac{\hbar^2}{2m}\nabla_2^2 - \frac{e^2}{r_{2_B}} \qquad (7.76)$$

The corresponding Schrödinger equation

$$H_0\psi(1,2) = E\psi(1,2) \qquad (7.77)$$

is easily solved. The ground state solution is the product of the ground state wave functions for the two hydrogen atoms, namely

$$\psi(1,2) = 1s_A(1)1s_B(2) \qquad (7.78)$$

where

$$1s_A(1) = \frac{1}{\sqrt{\pi a_0^3}}e^{-r_{1_A}/a_0} \quad , \quad 1s_B(2) = \frac{1}{\sqrt{\pi a_0^3}}e^{-r_{2_B}/a_0} \qquad (7.79)$$

we have

$$H_0 1s_A(1)1s_B(2) = 2E_H 1s_A(1)1s_B(2) \qquad (7.80)$$

that is the total energy is equal to that of two H–atoms.

As the atoms move close enough to each other to interact, one contribution to the energy comes from the Van der Waals attraction. Since it is a small term compared with the measured bond energy of H_2, we ignore it.

The next approximation to the interaction energy can be obtained by perturbation theory. We write the Hamiltonian in the form

$$H = H_0 + V_I \qquad (7.81)$$

where H_0 is the Hamiltonian appropriate for infinite R and V_I is the interaction operator

$$V_I = e^2 \left(-\frac{1}{r_{1_B}} - \frac{1}{r_{2_A}} + \frac{1}{r_{12}} + \frac{1}{R} \right) \qquad (7.82)$$

whose contribution goes to zero as R becomes large; $V_I \to 0$ as $R \to \infty$.

Approximating the true wave function as $\psi(1,2) = 1s_A(1)1s_B(2)$ we write

$$H 1s_A(1)1s_B(2) = 2E_H 1s_A(1)1s_B(2) + V_I 1s_A(1)1s_B(2) \qquad (7.83)$$

Multiplying both sides of Eq.(7.83) by $1s_A(1)1s_B(2)$ and integrating over the coordinates of electrons 1 and 2, we find

$$< 1s_A(1)1s_B(2)|H|1s_A(1)1s_B(2) >= \qquad (7.84)$$

$$< 1s_A(1)1s_B(2)|H_0|1s_A(1)1s_B(2) >$$

$$+ < 1s_A(1)1s_B(2)|V_I|1s_A(1)1s_B(2) >$$

$$= 2E_H + < 1s_A(1)1s_B(2)|V_I|1s_A(1)1s_B(2) >$$

The integral of the interaction potential V_I may be expressed explicitly as

$$< 1s_A(1)1s_B(2)|V_I|1s_A(1)1s_B(2) >= \qquad (7.85)$$

$$\frac{e^2}{R} < 1s_A(1)1s_B(2)|1s_A(1)1s_B(2) >$$

$$- < 1s_A(1)1s_B(2)|\frac{e^2}{r_{1_B}}|1s_A(1)1s_B(2) >$$

$$- < 1s_A(1)1s_B(2)|\frac{e^2}{r_{2_A}}|1s_A(1)1s_B(2) >$$

$$+ < 1s_A(1)1s_B(2)|\frac{e^2}{r_{12}}|1s_A(1)1s_B(2) >$$

$$= \frac{e^2}{R} - < 1s_A(1)|\frac{e^2}{r_{1_B}}|1s_A(1) > - < 1s_B(2)|\frac{e^2}{r_{2_A}}|1s_B(2) >$$

$$+ < 1s_A(1)1s_B(2)|\frac{e^2}{r_{12}}|1s_A(1)1s_B(2) >$$

$$< V_I >= \frac{e^2}{R} - < \frac{e^2}{r_{1_B}} >_A - < \frac{e^2}{r_{2_A}} >_B + < \frac{e^2}{r_{12}} >_{AB} \qquad (7.86)$$

$$E =< H >= 2E_H+ < V_I > \qquad (7.87)$$

$$\Delta E =< H > -2E_H =< V_I > \qquad (7.88)$$

In general, for a diatomic molecule AB,

$$< V_I(R) >= E_{AB}(R) - E_A - E_B = \Delta E(R) \qquad (7.89)$$

Eq.(7.85) provides an expression for the interaction energy ΔE, the difference in energy between the molecule described by the wave function $1s_A(1)1s_B(2)$ and the separated atoms. If we evaluate the interaction energy for different values of the internuclear distance R, we obtain the result shown in the figure.

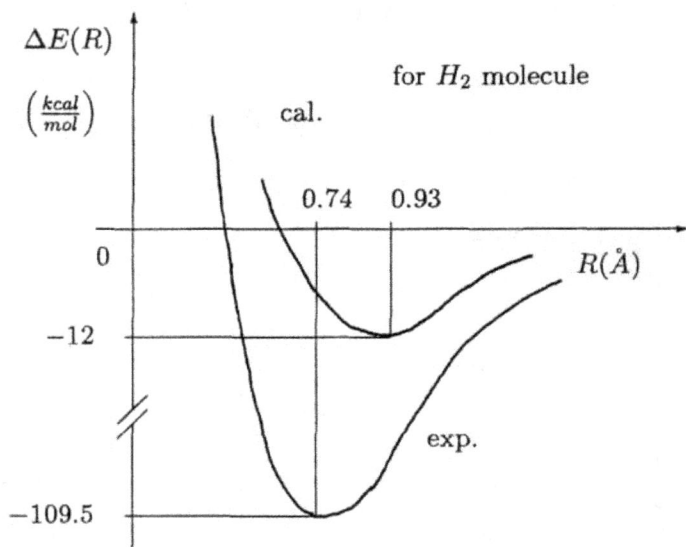

Fig. 7.7. Experimental and calculated interaction energy.

The wave function $1s_A(1)1s_B(2)$ yields only about 10% of the dissociation energy shows that it does not provide an adequate description of the hydrogen molecule. We must use a wave function that takes account of the indistinguishability of the electrons.

7.8   Valence–bond approach

Since the two electrons of the interacting hydrogen atoms are indistinguishable, we can build the required indistinguishability into the approximate wave function by writing

$$\psi_\pm(1,2) = \frac{1}{\sqrt{2}} \left[1s_A(1)1s_B(2) \pm 1s_B(1)1s_A(2)\right] \qquad (7.90)$$

One can calculate the interaction energy ΔE from $\psi_\pm$ by the method used previously, $\Delta E = < V_I >$.

$$\Delta E_\pm = < 1s_A(1)1s_A(1)|V_I|1s_B(2)1s_B(2) > \qquad (coulomb\ integral,\ Q)$$
$$(7.91)$$

$$\pm < 1s_A(1)1s_B(1)|V_I|1s_A(2)1s_B(2) > \quad (exchange\ integral,\ J)$$

$$\Delta E_\pm = Q \pm J \quad (7.92)$$

(E_+ is the state with lower energy) where $\Delta E_\pm$, Q, and J are all functions of R.

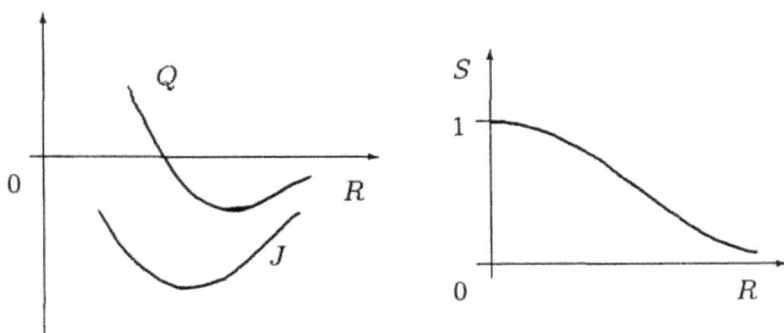

Fig. 7.8. Variation of Q, J, and S versus R.

The integral of the product of the functions $1s_A$ and $1s_B$, usually designated by S, and is called the **overlap integral**

$$S =< 1s_A(1)|1s_B(1) >=< 1s_A(2)|1s_B(2) > \quad (7.93)$$

S is also a function of R.

When S is included in the formulation, the normalized wave function is

$$\psi_\pm(1,2) = \frac{1}{\sqrt{2(1 \pm S^2}} [1s_A(1)1s_B(2) \pm 1s_B(1)1s_A(2)] \quad (7.94)$$

The expression for $\Delta E_\pm$ becomes

$$\Delta E_\pm = \frac{Q \pm J}{1 \pm S^2} \quad (7.95)$$

ΔE_+ : bonding ground state , ΔE_- : excited repulsive state.

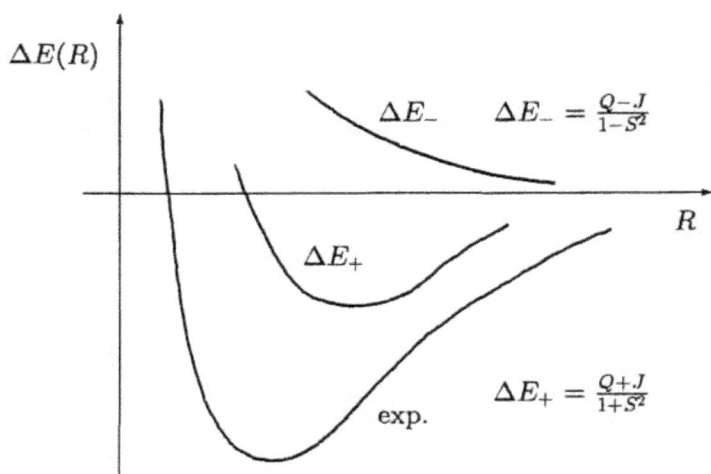

Fig. 7.9. Bonding and antibonding energy levels.

This treatment of the covalent bond, first worked out by Heitler and London for H_2, and further developed by Slater and Pauling, is based upon a bonding wave function for the two electrons which are shared between the two atoms. It is often called the **valence–bond approach**. The bonding function ψ_+ corresponds to a state with paired electron spins, while the repulsive function ψ_- has parallel spins; that is, the bonding function is a singlet and the antibonding function is a triplet. ψ_+ (singlet) , ψ_- (triplet).

The probability densities for electrons 1 and 2 are

$$P_\pm(1) = \int |\psi_\pm(1,2)|^2 dv_2 \tag{7.96}$$

$$= \frac{1}{2(1 \pm S^2)} \left[1s_A(1)^2 + 1s_B(1)^2 \pm 2S \cdot 1s_A(1)1s_B(1) \right]$$

$$P_\pm(2) = \frac{1}{2(1 \pm S^2)} \left[1s_A(2)^2 + 1s_B(2)^2 \pm 2S \cdot 1s_A(2)1s_B(2) \right] \tag{7.97}$$

$$P_\pm(1) = P_\pm(2) \tag{7.98}$$

The total charge density

$$\rho_\pm = e\left[P_\pm(1) = P_\pm(2) \right] = 2eP_\pm \tag{7.99}$$

$$\rho_{\pm}(x, y, z) = \frac{e}{\pi a_0^2 (1 \pm S^2)} \left[e^{-2r_A/a_0} + e^{-2r_B/a_0} \pm 2S \cdot e^{-(r_A+r_B)/a_0} \right]$$

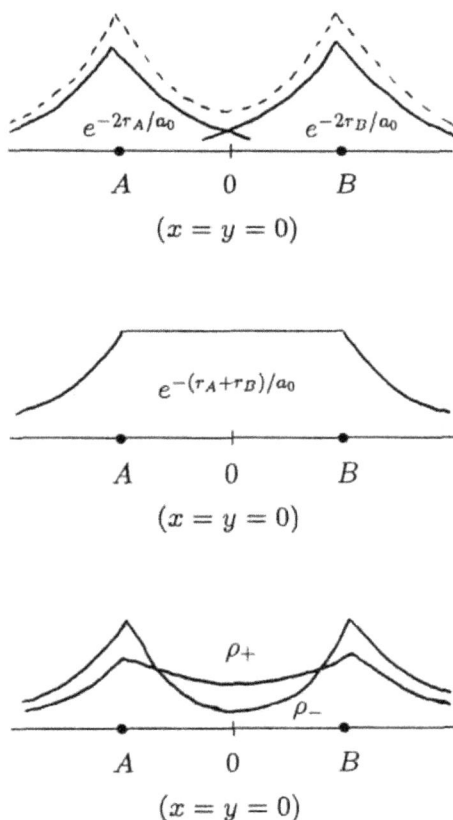

$$(7.100)$$

$$e^{-2r_A/a_0} \qquad e^{-2r_B/a_0}$$

$$A \qquad 0 \qquad B$$

$$(x = y = 0)$$

$$e^{-(r_A+r_B)/a_0}$$

$$A \qquad 0 \qquad B$$

$$(x = y = 0)$$

$$\rho_+$$

$$\rho_-$$

$$A \qquad 0 \qquad B$$

$$(x = y = 0)$$

Fig. 7.10. Charge density distributions.

If the total energy E and its variation with R is known, we can use the virial theorem to calculate the average kinetic energy $< T >$ and the average potential energy $< V >$.

For a diatomic molecule, the virial theorem is

$$< T >= -\frac{1}{2} < V > -\frac{1}{2} R \frac{dE}{dR} \qquad (7.101)$$

at equilibrium $(R = R_e)$ $dE/dR = 0$ and then

$$< T >= -\frac{1}{2} < V > \tag{7.102}$$

Since

$$E =< T > + < V > \tag{7.103}$$

we can write

$$< T >= -\left(E + R\frac{dE}{dR}\right) \tag{7.104}$$

$$< V >= 2E + R\frac{dE}{dR} \tag{7.105}$$

7.9   Molecular–orbital approach

Molecular–orbital (MO) approach, developed by Hund and Mulliken, is a generalization of the atomic orbital approach to the electronic structure of atoms; that is, one builds up the configuration of a many–electron molecule by assigning electrons to one–electron wave functions or orbitals.

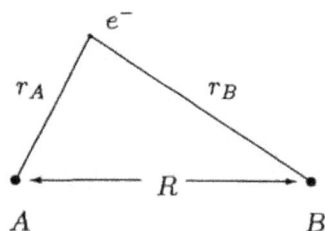

Fig. 7.11. Relative positions of nuclei and electron in H_2^+ ion.

The one–electron hydrogen molecule ion H_2^+ is the prototype for MO theory. The Hamiltonian for this one–electron system is

$$H = -\frac{\hbar^2}{2m}\nabla^2 - \frac{e^2}{r_A} - \frac{e^2}{r_B} + -\frac{e^2}{R} \tag{7.106}$$

The lack of spherical symmetry makes the molecular problem considerably more difficult.

The wave functions of H_2^+ are expected to have large amplitude in the vicinity of the nuclei, where the electron–nuclear attraction is large. In addition, the ground state orbital will be nodeless, while the first excited state will have a single node, and so on.

A reasonable representation of the approximately normalized, nodeless, ground state orbital is

$$\phi_b = \frac{1}{\sqrt{2}}(1s_A + 1s_B) = \phi_+ \tag{7.107}$$

This is the simplest example of the method of **linear combination of atomic orbitals** (LCAO) for representing molecular orbitals. The overlap of the two functions has been neglected.

We can write the approximate first excited MO

$$\phi_a = \frac{1}{\sqrt{2}}(1s_A - 1s_B) = \phi_- \tag{7.108}$$

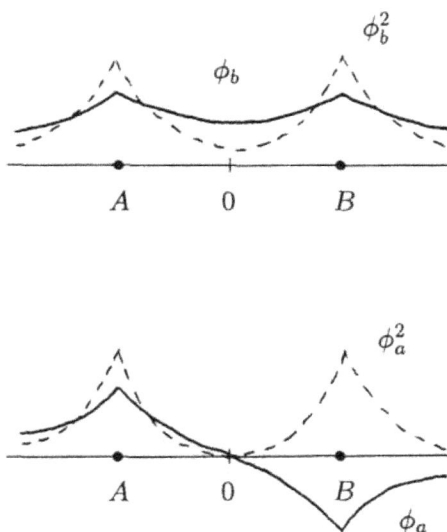

Fig. 7.12. Bonding and antibonding molecular orbitals.

Defining the effective one–electron Hamiltonian:

$$h_e(1) = -\frac{\hbar^2}{2m}\nabla_1^2 - \frac{e^2}{r_{1_A}} - \frac{e^2}{r_{1_B}} \qquad (7.109)$$

$$E_b = <\phi_b|h_e(1)|\phi_b> \qquad (7.110)$$

$$= \frac{1}{2}[<1s_A|h|1s_A> + <1s_B|h|1s_B>]$$

$$+ \frac{1}{2}[<1s_A|h|1s_B> + <1s_B|h|1s_A>]$$

$$= <1s_A|h|1s_A> + <1s_A|h|1s_B> = \alpha + \beta$$

$$<1s_A|h|1s_A> = <1s_B|h|1s_B> \text{ and } <1s_A|h|1s_B> = <1s_B|h|1s_A>$$

$$\alpha = <1s_A|h|1s_A> \text{ atomic integral (or coulombic integral)}$$

$$\beta = <1s_A|h|1s_B> \text{ resonance integral}$$

Similarly

$$E_a = \alpha - \beta \qquad (7.111)$$

The interaction energy for the orbital ϕ_b is : $\Delta E_b = E_b - E_H \cong \beta$
The interaction energy for the orbital ϕ_a is : $\Delta E_a = E_a - E_H \cong -\beta$
$E_H \cong \alpha$

Since calculated values of β are found to be negative, $(\Delta E_b < 0)$, ϕ_b is a bonding MO, while, $(\Delta E_a > 0)$ ϕ_a is an antibonding MO.

Other bonding and antibonding MOs can be formed in a corresponding way by using excited atomic orbitals (e.g., $2s_A$, $2p_A$; $2s_B$, $2p_B$,...).

The atomic and resonance integrals α and β play a fundamental role in MO theory, analogous to that of the coulomb and exchange integrals Q and J in VB approach.

Introducing the correctly normalized functions

$$\phi_b = \frac{1}{\sqrt{2(1+S)}}(1s_A + 1s_B) \ , \quad \phi_a = \frac{1}{\sqrt{2(1+S)}}(1s_A - 1s_B) \qquad (7.112)$$

we obtain for the energies

$$E_b = \frac{\alpha + \beta}{1 + S} \quad , \quad E_a = \frac{\alpha - \beta}{1 - S} \tag{7.113}$$

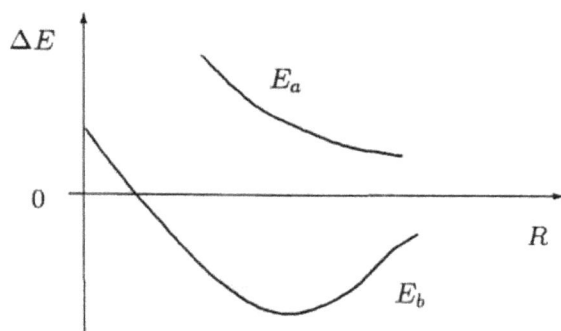

E_a: Antibonding (first excited state)
E_b: Bonding (ground state)

Fig. 7.13. Bonding and antibonding energy levels.

Some useful integrals are:

$$S(R) = < \phi_A | \phi_B > = e^{-R}(1 + R + \frac{R^2}{3}) \tag{7.114}$$

$$\varepsilon_{AA} = < \phi_A | - \frac{1}{r_B} | \phi_A > = -\frac{1}{R}[1 - e^{-2R}(1 + R)] \tag{7.115}$$

$$\varepsilon_{AB} = < \phi_A | - \frac{1}{r_B} | \phi_B > = -e^{-R}(1 + R) \tag{7.116}$$

To employ the simple MOs for a treatment of H_2, we make use of the Aufbau principle: The ground state of H_2 has the two electrons paired in the MO of lowest energy, ϕ_b,

$$\psi_1(1,2) = \phi_b(1)\phi_b(2) = \frac{1}{2}[1s_A(1) + 1s_B(1)][1s_A(2) + 1s_B(2)] \tag{7.117}$$

Since the two electrons are in the same orbital, the Pauli principle requires that they have paired spins; that is, the H_2 ground state in MO theory is a singlet.

The wave function must be antisymmetric with respect to electron interchange

$$\psi_1 = \frac{1}{\sqrt{2}} \begin{vmatrix} \phi_b\alpha(1) & \phi_b\beta(1) \\ \phi_b\alpha(2) & \phi_b\beta(2) \end{vmatrix} \tag{7.118}$$

$$= \phi_b(1)\phi_b(2)\frac{1}{\sqrt{2}}[\alpha(1)\beta(2) - \beta(1)\alpha(2)]$$

The state having the highest energy is formed by choosing the antibonding spin orbitals $\phi_a\alpha$ and $\phi_a\beta$ for occupancy:

$$\psi_2 = \frac{1}{\sqrt{2}} \begin{vmatrix} \phi_a\alpha(1) & \phi_a\beta(1) \\ \phi_a\alpha(2) & \phi_a\beta(2) \end{vmatrix} \tag{7.119}$$

$$= \phi_a(1)\phi_a(2)\frac{1}{\sqrt{2}}[\alpha(1)\beta(2) - \beta(1)\alpha(2)]$$

States of intermediate energy result when the orbitals ϕ_a and ϕ_b are selected for single occupancy:

$$\psi_3 = \frac{1}{\sqrt{2}}[\phi_a(1)\phi_b(2) + \phi_b(1)\phi_a(2)]\frac{1}{\sqrt{2}}[\alpha(1)\beta(2) - \beta(1)\alpha(2)] \tag{7.120}$$

$$\psi_4 = \frac{1}{\sqrt{2}}[\phi_a(1)\phi_b(2) + \phi_b(1)\phi_a(2)] \left\{ \begin{array}{l} \alpha(1)\alpha(2) \\ \frac{1}{\sqrt{2}}[\alpha(1)\beta(2) + \beta(1)\alpha(2)] \\ \beta(1)\beta(2) \end{array} \right\} \tag{7.121}$$

ψ_1, ψ_2, and ψ_3 are singlet states, and ψ_4 is a triplet state.

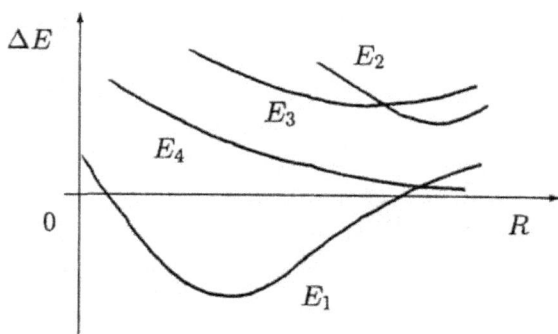

Fig. 7.14. Ground and excited electronic states of H_2.

The regions of the electromagnetic spectrum:

Radiowaves: spin orientation in magnetic field; change of spin		molecular rotations; change of orientation	molecular vibrations; change of configuration	UV,vis valence elect. trans.	Far-UV vac. UV; ioniz.	electron transitions of inner orbitals	nuclear reactions; change of nuclear configuration
				change of electron distribution			
NMR	ESR	Microwave	Infra-red	Vis. & UV		X-Ray	γ-Ray
10^{-2}		1	10^{2}	10^{4}		10^{6}	10^{8} $\nu(cm^{-1})$
10^{10}		10^{8}	10^{6}	10^{4}		10^{2}	1 $\lambda(\mathring{A})$
3×10^{8}		3×10^{10}	3×10^{12}	3×10^{14}		3×10^{16}	3×10^{18} $\nu(Hz)$
10^{-1}		10	10^{3}	10^{5}		10^{7}	10^{9} $E(J/mol)$

7.10   Worked examples

Example - 7.1 :

Apply the linear variation method to H_2^+ using a trial function of the form $\phi = c_A u_A + c_B u_B$ where $u_A = (1/\sqrt{\pi})e^{-r_A}$, $u_B = (1/\sqrt{\pi})e^{-r_B}$ to obtain approximate eigenfunctions and eigenvalues for the ground and first excited states. The coordinates are as shown in the figure.

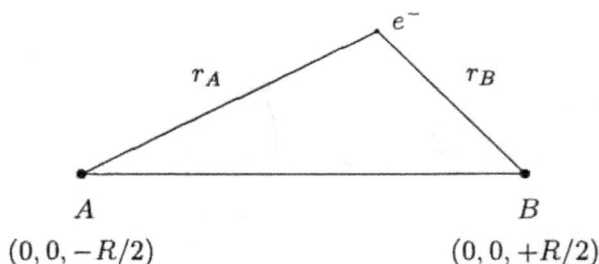

$$A$$
$$(0, 0, -R/2)$$

$$B$$
$$(0, 0, +R/2)$$

Solution :

The eigenvalues are obtained from the seqular equation

$$
\begin{vmatrix}
(H_{AA} - E) & (H_{AB} - E \cdot S) \\
(H_{BA} - E \cdot S) & (H_{BB} - E)
\end{vmatrix} = 0
$$

$H_{AA} = H_{BB} = \int u_A^* H u_A d\tau = < u_A|H|u_A >$
$H_{AB} = H_{BA} = < u_A|H|u_B >$, $S = < u_A|u_B > = < u_B|u_A >$
Therefore,
$(H_{AA} - E)^2 = (H_{AB} - E \cdot S)^2 \rightarrow (H_{AA} - E) = \pm(H_{AB} - E \cdot S)$

$$
E_\pm = \frac{H_{AA} \pm H_{AB}}{1 \pm S}
$$

The coefficients are obtained from the equation:
$(H_{AA} - E)C_A + (H_{AB} - E \cdot S)C_B = 0$ by substituting $E_\pm$ for E.
For E_+: $C_A = C_B$ and for E_-: $C_A = -C_B$
$C_A = \pm C_B$; $C_A + C_B = C_+$, $C_A - C_B = C_-$

For the magnitudes of the coefficients we use the normalization condition:

$$\int |\phi_\pm|^2 d\tau = |C_\pm|^2 \int (u_A^2 + u_B^2 \pm 2u_A u_B)d\tau = 1$$

or $\quad 2(1 \pm S)|C_\pm|^2 = 1 \quad$ or $\quad C_\pm = 1/\sqrt{2(1 \pm S)}$.

Therefore

$$< u_A|H|u_A > = < u_A| - \nabla^2 - \frac{2}{r_A} - \frac{2}{r_B} + \frac{2}{R}|u_A >$$

$$= E_H + \frac{2}{R} - < u_A|\frac{2}{r_B}|u_A >$$

$$(-\nabla^2 - \frac{2}{r_A})u_A = E_H u_A$$

For H_{AB} we find:

$$< u_A|H|u_B > = < u_A| - \nabla^2 - \frac{2}{r_B} - \frac{2}{r_A} + \frac{2}{R}|u_B >$$

$$= S(E_H + \frac{2}{R}) - < u_A|\frac{2}{r_A}|u_B >$$

$$E_\pm = E_H + \frac{2}{R} - \left[\frac{< u_A|\frac{2}{r_B}|u_A > \pm < u_A|\frac{2}{r_A}|u_B >}{(1 \pm S)}\right]$$

Example - 7.2 :

Two H-atoms having a large internuclear separation R interact through their instantaneous electric dipole moments. Show that the perturbation Hamiltonian which describes the interatomic interaction when $R \ll a_0$ can be written in the form: $H' = (e^2/R^3)(x_1 x_2 + y_1 y_2 - 2z_1 z_2)$. Here (x_1, y_1, z_1) and (x_2, y_2, z_2) are the coordinates of electrons 1 and 2 with respect to nuclei A and B, respectively.

Solution :

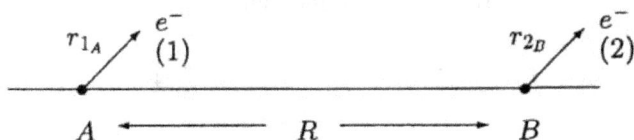

The complete Hamiltonian for the two H–atoms is $H = H_{0_A} + H_{0_B} + H'$ where

$$H' = e^2 \left(-\frac{1}{r_{1_B}} - \frac{1}{r_{2_A}} + \frac{1}{r_{12}} + \frac{1}{R} \right)$$

The position vectors can be written as:
$\mathbf{r}_{1_A} = x_1\mathbf{i} + y_1\mathbf{j} + z_1\mathbf{k}$; $\mathbf{r}_{2_B} = x_2\mathbf{i} + y_2\mathbf{j} + z_2\mathbf{k}$;
$\mathbf{r}_{2_A} = x_2\mathbf{i} + y_2\mathbf{j} + (z_2 + R)\mathbf{k}$; $\mathbf{r}_{1_B} = x_1\mathbf{i} + y_1\mathbf{j} + (z_1 - R)\mathbf{k}$.
Thus,

$$H' = e^2 \left[-\frac{1}{\sqrt{x_1^2 + y_1^2 + (z_1 - R)^2}} - \frac{1}{\sqrt{x_2^2 + y_2^2 + (z_2 + R)^2}} \right]$$

$$+ e^2 \left[\frac{1}{\sqrt{(x_2 - x_1)^2 + (y_2 - y_1)^2 + (z_2 - z_1 + R)^2}} + \frac{1}{R} \right]$$

using

$$(1 + x)^{-n} = 1 - nx + \frac{n(n+1)}{2!}x^2 + \dots ; \quad x^2 < 1$$

Therefore:

$$H' = \frac{e^2}{R} \left\{ -\left[1 + \frac{(r_{1_A}^2 - 2z_1 R)}{R^2} \right]^{-1/2} - \left[1 + \frac{(r_{2_B}^2 + 2z_2 R)}{R^2} \right]^{-1/2} \right\}$$

$$+ \frac{e^2}{R} \left\{ \left[1 + \frac{(r_{1_A}^2 + r_{2_B}^2 - 2(x_1 x_2 + y_1 y_2 + z_1 z_2) + 2R(z_2 - z_1))}{R^2} \right]^{-1/2} + 1 \right\}$$

$$H' = \frac{e^2}{R} \left\{ \frac{1}{R^2}(x_1 x_2 + y_1 y_2 + z_1 z_2) + \frac{3 \times 4R^2}{8R^4}(-2z_1 z_2) + \dots \right\}$$

$$H' = \frac{e^2}{R^3}(x_1 x_2 + y_1 y_2 - 2z_1 z_2)$$

This equation is in the form of the interaction between two electric dipole moments. It is often written as:

$$H' = \frac{e^2}{R^3}(\mathbf{r}_1 \cdot \mathbf{r}_2 - 3r_{1_z}r_{2_z})$$

Example - 7.3 :

Consider H_2^+ ion, and let R be the separation of the nuclei. Assume that the bonding MO can be approximated by a $1s$ AO $\psi(r) = Ce^{-\zeta r}$ with origin at the midpoint of the internuclear axis, C is a normalization constant. We will now perform a variational calculation of the ground state total energy of H_2^+, varying both ζ and R. It is convenient to use polar coordinates with the origin placed at the origin of $\psi(r)$.
a) Determine the total energy as a function of ζ and R for H_2^+ in the state $\psi(r)$.
b) Assume initially that R is 2.00 a.u. (the experimental equilibrium separation, R_e). Determine for $R = R_e$ the value of $\zeta(= \zeta_e)$ that minimizes the ground state total energy in (a). Calculate the total energy for $\zeta = \zeta_e$.
c) Let us now keep ζ fixed at $\zeta = \zeta_e$ and vary the internuclear separation, R. Determine for $\zeta = \zeta_e$ the equilibrium distance and the corresponding ground state energy using the energy expression from (a).
d) We can continue the iterative process initiated in (b) and (c) until a self–consistent solution is obtained. Since the total energies obtained in the two iterations deviate by 10^{-3} a.u. (verify that), we will consider the solutions obtained in (c) as the converged result. Use the result of (c) to calculate the spectroscopic dissociation energy for H_2^+.
e) The most accurate total energy and dissociation energy for H_2^+ are -0.6026 a.u. and 2.793 eV, respectively. What is the reason for the large discrepancy between the computed and experimental dissociation energies?

Solution :

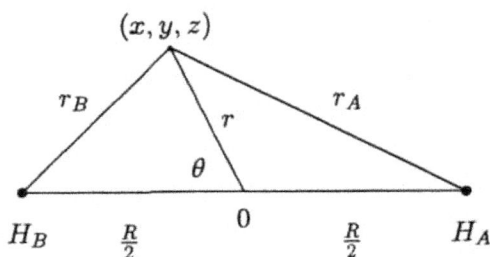

a)

$$\psi(r) = \left(\frac{\zeta^3}{\pi}\right)^{1/2} e^{-\zeta r} \quad , \quad H = -\frac{1}{2}\nabla^2 - \frac{1}{r_A} - \frac{1}{r_B} + \frac{1}{R}$$

$$< \psi(r)| -\frac{1}{2}\nabla^2 |\psi(r) > = \frac{1}{2}\zeta^2 \quad , \quad r_B = (\frac{R^2}{4} + r^2 - Rr\cos\theta)^{1/2}$$

$$< \psi(r)| -\frac{1}{r_B} |\psi(r) > = \frac{2\zeta^3}{R} \int_0^\infty r e^{-2\zeta r} dr \int_{r+R/2}^{|r-R/2|} \frac{du}{\sqrt{u}}$$

$$= e^{-\zeta R}(\zeta + \frac{2}{R}) - \frac{2}{R}$$

$$E(\zeta) = \frac{1}{2}\zeta^2 - \frac{3}{R} + 2e^{-\zeta R}(\zeta + \frac{2}{R})$$

b)

$$\frac{\partial E}{\partial \zeta} = 0 \;\rightarrow\; e^{-\zeta R} = \frac{\zeta}{2 + 2R\zeta}$$

at $R = R_e = 2.00$ a.u. we obtain $e^{-2\zeta} = \zeta/(2 + 4\zeta)$, which, solved numerically, within an accuracy of 10^{-4} a.u. yields $\zeta = \zeta_e = 0.912$ a.u. and $E(\zeta_e) = -0.467$ a.u. $\cong -12.71$ eV .

c) At $\zeta = \zeta_e = 0.912$ a.u. $\partial E/\partial R = 0$ gives

$$e^{\zeta_e R} = \frac{2}{3}[2 + 2\zeta_e R + \zeta_e^2 R^2]$$

which to an accuracy of 10^{-4} a.u. yields $R_e = 1.894$ a.u., $E(R_e) = -0.468$ a.u. $\Delta E = E(\zeta_e) - E(R_e) = -0.467 + 0.468 = 10^{-3}$ a.u..

d) $D_e = E(\infty) - E(R_e) = \zeta_e^2/2 - E(R_e) = 0.884$ a.u. $= 24.05$ eV.

e) The large difference is mainly caused by the fact that the trial function predicts an incorrect dissociation of H_2^+, namely $H^+ + H^+ + e^-$.

Example - 7.4 :

Assume that the total asymmetry of charge in a heteronuclear diatomic molecule is determined by the doubly occupied MO: $\psi = N(u_A + \lambda u_B)$ where u_A and u_B are pure AOs. A charge of $+e$ is centered on each nucleus and the internuclear distance is R. Derive an expression for the electric dipole moment μ in terms of λ, R, and the overlap integral S. You may neglect the integral $< u_A|z|u_B >$ where the bond lies along the z-direction if the origin is taken at the center point of the bond.

Solution :

$$A \qquad\qquad\qquad\qquad\qquad\qquad\qquad\qquad\qquad B$$
$$\bullet\hspace{3cm}|\hspace{6cm}\bullet$$
$$-R/2 \qquad\qquad\qquad\qquad\qquad\qquad\qquad +R/2$$

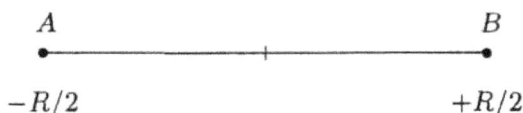

$$\int \psi^* \psi d\tau = N^2 \int (u_A^2 + \lambda^2 u_B^2 + 2\lambda u_A u_B) d\tau = N^2(1 + \lambda^2 + 2\lambda S) = 1$$

Therefore $N = (1 + \lambda^2 + 2\lambda S)^{-1/2}$

The dipole moment is given by $\mu = \sum_{i=A,B} q_i z_i + \int \psi^*(-2ez)\psi d\tau$

$$\mu = -2eN^2 \int (zu_A^2 + z\lambda^2 u_B^2 + 2z\lambda u_A u_B) d\tau$$

$$= -2eN^2(< u_A|z|u_A > +\lambda^2 < u_B|z|u_B > +2\lambda < u_A|z|u_B >)$$

Since u_A and u_B are pure AOs, we can replace $< u_A|z|u_A >$ and $< u_B|z|u_B >$ with $-R/2$ and $+R/2$, respectively. The result is

$$\mu = \frac{eR(1 - \lambda^2)}{1 + \lambda^2 + 2\lambda S}$$

Example - 7.5 :

Obtain analytical expressions for the following molecular integrals in terms of internuclear separation R:

a) $S = <u_A|u_B> = \frac{1}{2} \int e^{-(r_A+r_B)} d\tau$,

b) $I = <u_A|\frac{2}{r_D}|u_A> = \frac{1}{\pi} \int e^{-r_A} \frac{2}{r_D} e^{-r_A} d\tau$.

Use the geometry given in Example–7.3 .

Solution :

This type of integrals can be evaluated easily by using elliptical coordinates. The necessary relations between the coordinates are:

$$\mu = (r_A + r_B)/R \quad 1 \le \mu \le \infty$$

$$\nu = (r_A - r_B)/R \quad -1 \le \mu \le +1$$

$$\phi \quad 0 \le \phi \le 2\pi$$

and the volume element is

$$d\tau = \frac{R^3}{8}(\mu^2 - \nu^2)d\mu d\nu d\phi$$

a)

$$S = \frac{1}{\pi} \int_0^{2\pi} \int_{-1}^{+1} \int_{+1}^{\infty} e^{-\mu R} \frac{R^3}{8}(\mu^2 - \nu^2)d\mu d\nu d\phi$$

$$= e^{-R}(1 + R + \frac{R^2}{3})$$

b)

$$I = \frac{1}{\pi} \int_0^{2\pi} \int_{-1}^{+1} \int_{+1}^{\infty} e^{-R(\mu+\nu)} \frac{2 \cdot 2}{R(\mu - \nu)} \frac{R^3}{8}(\mu^2 - \nu^2)d\mu d\nu d\phi$$

$$= \frac{2}{R}[1 - e^{-2R}(1 + R)].$$

Chapter 8

Approximation Methods for Many–Electron Systems

$$V(\mathbf{r}_1, \mathbf{r}_2, ..., \mathbf{r}_N) \cong \sum_{i=1}^{N} V_{eff}(\mathbf{r}_i)$$

8.1   Many–electron system calculations

Schrödinger's equation can not be solved exactly for two–electron atoms, so that certain approximation methods must be used. Accurate results can be obtained for the energy levels and wave functions of two–electron atom by performing variational calculations. But this method becomes increasingly tedious when the number of electrons increases in the atomic system. Therefore some other general methods should be developed to study many–electron atomic and molecular systems.

8.2   The Thomas–Fermi model of the atoms

The Thomas–Fermi (TF) model is based on statistical and semiclassical considerations for the ground state of many–electron atoms. In this model the N electrons of the system are treated as a Fermi electron gas in the ground state, confined to a region of space by a central potential $V(r)$ which vanishes at infinity.

The aim of the TF model is to provide a method of calculating the potential $V(r)$ and the electron density $\rho(r)$.

The total energy of the system is written as

$$E = E_F + V(r) \tag{8.1}$$

In the Fermi electron gas model E_F is given as

$$E_F = \frac{\hbar^2}{2m} \left(3\pi^2 \rho\right)^{2/3} \tag{8.2}$$

Combining Eqs.(8.1) and (8.2), we have

$$\rho(r) = \frac{1}{3\pi^2} \left(\frac{2m}{\hbar^2}\right)^{3/2} (E - V(r))^{3/2} \tag{8.3}$$

$\rho = 0$ for $V \geq E$. Taking the relation between the electrostatic potential $\phi(r)$ and the potential energy $V(r)$

$$\phi(r) = -\frac{1}{e} V(r) \tag{8.4}$$

and

$$\phi_0 = -\frac{1}{e}E \tag{8.5}$$

one may define

$$\Phi(r) = \phi(r) - \phi_0 \tag{8.6}$$

Thus,

$$\rho(r) = \frac{1}{3\pi^2}\left(\frac{2m}{\hbar^2}\right)^{3/2}(e\Phi(r))^{3/2} \quad for \quad \Phi \geq 0 \tag{8.7}$$

$$\rho(r) = 0 \quad for \quad \Phi < 0$$

The Poisson's equation for a given charge density is

$$\nabla^2\Phi(r) = e\rho(r) \tag{8.8}$$

Substituting Eq.(8.7) in Eq.(8.8), we have

$$\nabla^2\Phi(r) = \frac{e}{3\pi^2}\left(\frac{2m}{\hbar^2}\right)^{3/2}(e\Phi(r))^{3/2} \quad for \quad \Phi \geq 0 \tag{8.9}$$

$$\nabla^2\Phi(r) = 0 \quad for \quad \Phi < 0$$

Defining

$$\chi = \frac{1}{Ze}r\Phi(r) \tag{8.10}$$

and $x = r/b$ where

$$b = \left(\frac{(3\pi)^{2/3}}{2^{7/3}}\right)a_0 Z^{-1/3}$$

charge density may be expressed as

$$\rho = \frac{Z}{b^3}\left(\frac{\chi}{x}\right)^{3/2} \quad for \quad \chi \geq 0 \tag{8.11}$$

$$\rho = 0 \quad for \quad \chi < 0$$

Knowing that $\nabla^2\Phi \rightarrow (1/r)d^2[r\Phi(r)]/dr^2$; then one can write

$$\frac{d^2\chi}{dx^2} = \frac{1}{\sqrt{x}}\chi^{3/2} \tag{8.12}$$

This equation is known as the **Thomas–Fermi equation** (it is usually solved numerically)

$d^2\chi/dx^2 = 0$ for $\chi < 0$ and $\chi(0) = 1$. $\chi(\infty) = 0$ for neutral atom solution.

Knowing the function $\chi(x)$, we can obtain the function $\Phi(r)$, and hence the electrostatic potential $\phi(r)$, the potential energy $V(r)$ and the density $\rho(r)$.

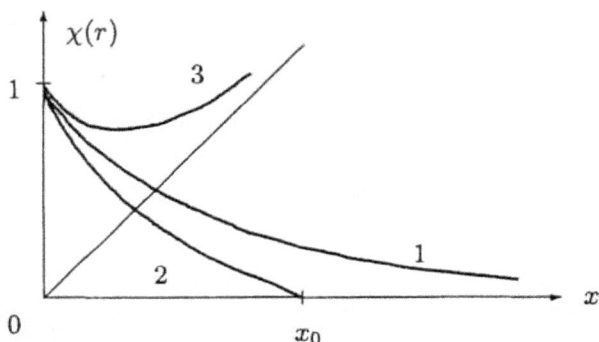

Fig. 8.1. Solutions of the Thomas–Fermi equation.

(1): Neutral atom solution; a solution which is asymptotic to the x–axis.
(2): Solution for positive ion $(N < Z)$; solutions which vanish for a finite values $x = x_0$.
(3): Solution for a neutral atom under pressure; solutions which never vanish and diverge for large x.

In the TF model the central potential $V(r)$ is given for neutral atoms by

$$V(r) = -\frac{Ze^2}{r}\chi \quad in \ SI \ units \quad V(r) = -\frac{Ze^2}{(4\pi\varepsilon_0)r}\chi \tag{8.13}$$

where

$$\chi(x) \cong 1 - 1.588x + ... \tag{8.14}$$

or

$$V(r) \cong e^2 \left(-\frac{Z}{r} + 1.794 \frac{Z^{4/3}}{a_0} \right) \tag{8.15}$$

The first term on the rihgt–hand–side of this equation represents the nuclear attraction, and the second term represents the electron repulsion. In *SI* units:

$$V(r) \cong \frac{e^2}{4\pi\varepsilon_0} \left(-\frac{Z}{r} + 1.794 \frac{Z^{4/3}}{a_0} \right) \tag{8.16}$$

In *a.u.* :

$$V(r) \cong -\frac{Z}{r} + 1.794 Z^{4/3} \tag{8.17}$$

8.3 Hartree–SCF method

In this method the energy and wave function of a many–electron atom is calculated numerically. Consider an N–electron atom with atomic number Z. The one–electron Hamiltonian for electron i is (in *a.u.*)

$$H_i^{(0)} = -\frac{1}{2}\nabla_i^2 - \frac{Z}{r_i} \tag{8.18}$$

Let

$$H^{(0)} = \sum_{i=1}^{N} H_i^{(0)} \tag{8.19}$$

Total Hamiltonian (non–relativistic) is

$$H = H^{(0)} + \sum_{i<j} \frac{1}{r_{ij}} \tag{8.20}$$

Total wave function is assumed as a Hartree–product

$$\psi(1, 2, ..., N) = \prod_{i=1}^{N} \phi_i(i) \tag{8.21}$$

The atomic orbitals $\phi_i(i)$ are not assumed to be the hydrogen–like orbitals, they are unknown.

Define

$$V_i^{eff} = \sum_{j \neq i}^{N} <\phi_j(j)| \frac{1}{r_{ij}} |\phi_j(j)> \qquad (8.22)$$

The Schrödinger equation (the eigenvalue equation) for electron i is

$$\left(H_i^{(0)} + V_i^{eff} \right) \phi_i(i) = \varepsilon_i \phi_i(i) \qquad (8.23)$$

This equation is called as the **Hartree–equation**.

Define the Coulomb operator

$$J_j(i) = <\phi_j(j)| \frac{1}{r_{ij}} |\phi_j(j)> \qquad (8.24)$$

Coulomb operator is an integral operator.

$$V_i^{eff} = \sum_{j \neq i}^{N} J_j(i) \qquad (8.25)$$

We rewrite Eq.(8.23) as

$$\left(H_i^{(0)} + \sum_{j \neq i} J_j(i) \right) \phi_i(i) = \varepsilon_i \phi_i(i) \qquad (8.26)$$

Eq.(8.26) is an integrodifferential equation. There are N such equations, one for each electron.

V_i^{eff} may be interpreted as: The ith electron is assumed to move in a potential field due to the average charge distribution of the other $(N - 1)$ electrons.

The difficulty in solving the N equations, Eq.(8.26), comes from V_i^{eff}, because V_i^{eff} depend upon the unknown AOs, $\{\phi_i(i)\}, i = 1, 2, ..., N$. Hartree showed that these orbitals could be obtained by an iterative technique.

As a first approximation one assumes that

$$J_j^{(0)}(i) = <\phi_j^{(0)}(j)| \frac{1}{r_{ij}} |\phi_j^{(0)}(j)> \qquad (8.27)$$

Here $\{\phi_j^{(0)}(j)\}$ are a suitable set of approximate orbitals.

One then defines the first approximation to the effective Hamiltonian (called the **Hartree operator**) by

$$H_i^{(1)} = H_i^{(0)} + \sum_{j \neq i} J_j^{(0)}(i) \tag{8.28}$$

and solving the N equations

$$H_i^{(1)} \phi_i^{(1)}(i) = \varepsilon_i^{(1)} \phi_i^{(1)}(i) \tag{8.29}$$

to obtain a first–improved set of orbitals $\{\phi_i^{(1)}(i)\}$. Eq.(8.29) is a pseudoeigenvalue equation (operator depends on the solution)

$$J_j^{(1)}(i) = < \phi_j^{(1)}(j)| \frac{1}{r_{ij}} |\phi_j^{(1)}(j) > \tag{8.30}$$

$$H_i^{(2)} = H_i^{(0)} + \sum_{j \neq i} J_j^{(1)}(i) \tag{8.31}$$

$$H_i^{(2)} \phi_i^{(2)}(i) = \varepsilon_i^{(2)} \phi_i^{(2)}(i) \tag{8.32}$$

The process is repeated until some set of coulombic operators $\{J_j^{(n)}(i)\}$ leads to Hartree operator such that (for all values of i)

$$H_i^{(n)} \cong H_i^{(n+1)} = H^{SCF} \quad , \quad \phi_i^{(n)} \cong \phi_i^{(n+1)} = \phi_i(i) \tag{8.33}$$

At this point the orbitals are such that the charge distribution arising from them will reproduce the field. One then says that the electrons move in a SCF. This method is known as the **Hartree–SCF method**. The final orbitals are usually referred to as **SCF AOs**.

The total energy of an N–electron atom in the hartree–SCF approximation is given by the expectation value of the correct Hamiltonian (not the SCF Hamiltonian). The role of the SCF Hamiltonian is just to generate the SCF AOs.

The total energy is then

$$< H >=< H^{(0)} > + \sum_{i<j} < \frac{1}{r_{ij}} > \tag{8.34}$$

If the AOs are normalized

$$< H^{(0)} > = \sum_{i=1}^{N} < \phi_i(i)|H_i^{(0)}|\phi_i(i) > = \sum_{i=1}^{N} \varepsilon_i^{(0)} \qquad (8.35)$$

$\varepsilon_i^{(0)}$ is the energy of a single electron i moving in some effective potential field provided by the nucleus and the other $N-1$ electrons.

The second part of the total energy is given by

$$\sum_{i<j} < \frac{1}{r_{ij}} > = \sum_{i<j} < \phi_i(i)\phi_j(j)|\frac{1}{r_{ij}}|\phi_i(i)\phi_j(j) > \qquad (8.36)$$

$$= \sum_{i<j} < \phi_i(i)|J_j(i)|\phi_i(i) > = \sum_{i<j} < \phi_j(j)|J_i(j)|\phi_j(j) > = \sum_{i<j} J_{ij}$$

The total energy given by the Hartree–SCF method is

$$E = \sum_{i=1}^{N} \varepsilon_i^{(0)} + \sum_{i<j}^{N} J_{ij} \qquad (8.37)$$

For He atom $E = -2.86$ $a.u.$

Hartree–SCF solutions satisfy the Virial theorem. The Hartree eigenvalue equation for the kth electron is

$$\left(H_k^{(0)} + \sum_{j \neq k} J_j(k) \right) \phi_k(k) = \varepsilon_k \phi_k(k) \qquad (8.38)$$

If the kth orbital is the highest occupied orbital, then $-\varepsilon_k$ is just the first ionization potential of the atom.

8.4 Hartree–Fock method

In the Hartree–SCF method it was assumed that each electron in an N–electron system moved in a potential field which was an average of that provided by the remaining $N-1$ electrons. The total electronic wave function, however, was not chosen to be antisymmetric, and thus the Pauli principle was not satisfied. If one uses the antisymmetrized Hartree product wave function and applying the variational principle one obtaines a new set of self–consistent–field equations, which are called as the **Hartree–Fock SCF equations.**

HF equations for closed–shell atoms:

Total non–relativistic Hamiltonian of an atom with $2N$ electrons is (in *a.u.*)

$$H = \sum_{\mu=1} 2N h_\mu + \sum_{\mu<\nu} \frac{1}{r_{\mu\nu}} \tag{8.39}$$

where

$$h_\mu = -\frac{1}{2}\nabla_\mu^2 - \frac{Z}{r_\mu} \tag{8.40}$$

is the one–electron Hamiltonian.

Consider the $2N$–electron wave function is constructed as a product of $2N$ spin orbitals

$$\Phi = \prod_{\mu=1}^{2N} S_\mu(\mu) \quad (Hartree\ product) \tag{8.41}$$

$$\Phi = S_1(1)S_2(2)\cdots S_{2N}(2N) \tag{8.42}$$

Spin orbitals:

$$S_\mu(\mu) \ \rightarrow \ \phi_i(\mu)\omega(\mu) \tag{8.43}$$

$\phi_i(\mu)$: Spatial orbital ; $\omega(\mu)$: Spin function.

$$\Phi = \phi_1(1)\alpha(1)\phi_1(2)\beta(2)\phi_2(3)\alpha(3)\phi_2(4)\beta(4) \tag{8.44}$$

$$\cdots\phi_N(2N-1)\alpha(2N-1)\phi_N(2N)\beta(2N)$$

We assume that the spatial orbitals are orthonormal,

$$< \phi_i(\mu)|\phi_j(\mu) >= \delta_{ij} \tag{8.45}$$

The expectation value of total Hamiltonian, Eq.(8.39), with the anti-symmetrized wave function may be written as

$$< H >=< \Phi|H|\Phi >=< \Phi_H|\mathcal{A}H\mathcal{A}|\Phi_H >=< \Phi_H|H\mathcal{A}^2|\Phi_H > \tag{8.46}$$

$$= \sqrt{N!} < \Phi_H|H\mathcal{A}|\Phi_H > = \sum_P (-1)^p < \Phi_H|H|\Phi_H > =$$

$$\sum_P (-1)^p < S_1(1) \cdots S_{2N}(2N)| \sum_\mu h_\mu + \sum_{\mu < \nu} \frac{1}{r_{\mu\nu}} |PS_1(1) \cdots S_{2N}(2N) >$$

$$(8.47)$$

Here we have taken $\Phi = \mathcal{A}\Phi_H$, $\mathcal{A}^2 = \sqrt{N!}\mathcal{A}$, $\mathcal{A} = (1/\sqrt{N!})\sum_P(-1)^pP$, and $\Phi_H \equiv$ Hartree product wave function. The summations over the identity permutations for the one–electron part of Eq.(8.47) may be written as

$$\sum_{\mu=1}^{2N} < h_\mu > = \sum_{\mu=1}^{2N} < S_\mu(\mu)|h_\mu|S_\mu(\mu) > \qquad (8.48)$$

Because of the assumed form of the wave function, the successive odd– and even–subscripted spin orbitals have the forms

$$S_\mu(\mu) = \phi_i(\mu)\alpha(\mu) \quad , \quad S_{\mu+1}(\mu+1) = \phi_i(\mu+1)\beta(\mu+1) \qquad (8.49)$$

Thus, after integration over spin functions Eq.(8.48) takes the form

$$2\sum_{i=1}^{N} < \phi_i(\mu)|h_\mu|\phi_i(\mu) > = 2\sum_{i=1}^{N} \varepsilon_i^{(0)} \qquad (8.50)$$

Therefore

$$\sum_{\mu=1}^{2N} < h_\mu > = 2\sum_{i=1}^{N} \varepsilon_i^{(0)} \qquad (8.51)$$

Eq.(8.51) represents the zeroth order energy of the system which corresponds to the independent–particle model result.

Two–electron portion of $< H >$:

For identity permutations we obtain

$$\sum_{\mu < \nu} < S_\mu(\mu)S_\nu(\nu)|\frac{1}{r_{\mu\nu}}|S_\mu(\mu)S_\nu(\nu) > \qquad (8.52)$$

Now whenever μ is odd and $\nu = \mu + 1$ we have

$$S_\mu(\mu) = \phi_i(\mu)\alpha(\mu) \quad , \quad S_\nu(\nu) = \phi_i(\nu)\beta(\nu) \qquad (8.53)$$

There are N such occurences, so this gives

$$\sum_{i=1}^{N} < \phi_i(\mu)\phi_i(\nu)|\frac{1}{r_{\mu\nu}}|\phi_i(\mu)\phi_i(\nu) > = \sum_{i=1}^{N} J_{ij} \qquad (8.54)$$

The general expression for the Coulombic integral is

$$J_{ij} = < \phi_i(\mu)\phi_j(\nu)|\frac{1}{r_{\mu\nu}}|\phi_i(\mu)\phi_j(\nu) > \qquad (8.55)$$

For all other values of $\mu < \nu$ there are **four** ways one can obtain a given Coulombic integral J_{ij}.

$$S_\mu(\mu) = \left\{ \begin{array}{c} \phi_i(\mu)\alpha(\mu) \\ \phi_i(\mu)\beta(\mu) \end{array} \right\} \quad , \quad S_\nu(\nu) = \left\{ \begin{array}{c} \phi_j(\nu)\alpha(\nu) \\ \phi_j(\nu)\beta(\nu) \end{array} \right\} \qquad (8.56)$$

Identity permutations then lead to a two–electron contribution:

$$\sum_{i=1}^{N} J_{ij} + 4 \sum_{i<j}^{N} J_{ij} \qquad (8.57)$$

In the case of two–electron permutations (odd parity) one obtaines

$$- \sum_{\mu<\nu} < S_\mu(\mu)S_\nu(\nu)|\frac{1}{r_{\mu\nu}}|S_\nu(\mu)S_\mu(\nu) > \qquad (8.58)$$

when μ is odd and $\nu = \mu + 1$ this leads to zero integrals as a result of spin orthogonality. For the other values of $\mu < \nu$ one obtains:

$$< \phi_i(\mu)\phi_j(\nu)|\frac{1}{r_{\mu\nu}}|\phi_j(\mu)\phi_i(\nu) > = K_{ij} \qquad (8.59)$$

where K_{ij} are called exchange integrals. Of the four $\mu < \nu$ combinations, only two lead to non–zero integrals for two–electron permutations. Thus, the total contribution due to two–electron permutations is

$$2 \sum_{i<j}^{N} K_{ij} \qquad (8.60)$$

Permutations involving more than two electrons lead to zero integral. Therefore,

$$< H >= 2 \sum_{i=1}^{N} \varepsilon_i^{(0)} + \sum_{i=1}^{N} J_{ii} + \sum_{i<j}^{N} (4J_{ij} - 2K_{ij}) \qquad (8.61)$$

Considering

$$J_{ij} = J_{ji} \quad , \quad K_{ij} = K_{ji} \quad , \quad J_{ii} = K_{ii} \qquad (8.62)$$

one can write

$$< H >= 2 \sum_{i=1}^{N} \varepsilon_i^{(0)} + \sum_{i=1}^{N} J_{ii} + 2 \sum_{i<j}^{N} (2J_{ij} - K_{ij}) \qquad (8.63)$$

$$= 2 \sum_{i=1}^{N} \varepsilon_i^{(0)} + \sum_{i,j}^{N} (2J_{ij} - K_{ij})$$

This equation, Eq.(8.63), is valid only for closed–shell atoms in which the total wave function is approximated as a single determinant of doubly occupied spatial orbitals.

We now wish to find the best possible orbitals to use in a wave function restricted to the single–determinantal form. Using the variational method, we consider the functional

$$J = 2 \sum_{i=1}^{N} \varepsilon_i^{(0)} + \sum_{i,j}^{N} (2J_{ij} - K_{ij}) - \sum_{i,j}^{N} \lambda_{ij} (< \phi_i | \phi_j > -\delta_{ij}) \qquad (8.64)$$

where λ_{ij} are Langrangian multipliers. We require that $\delta J = 0$ for small variations $\delta \phi_i$ of these optimum orbitals. Define the Coulomb and exchange operators, respectively,

$$J_i(\mu)\phi_j(\mu) =< \phi_i(\nu)|\frac{1}{r_{\mu\nu}}|\phi_i(\nu) > \phi_j(\mu) \qquad (8.65)$$

$$K_i(\mu)\phi_j(\mu) =< \phi_i(\nu)|\frac{1}{r_{\mu\nu}}|\phi_j(\nu) > \phi_i(\mu) \qquad (8.66)$$

Then the Coulomb and exchange integrals can be rewritten as

$$J_{ij} =< \phi_i(\mu)|J_j(\mu)|\phi_i(\mu) >=< \phi_j(\nu)|J_i(\nu)|\phi_j(\nu) > \qquad (8.67)$$

$$K_{ij} = <\phi_i(\mu)|K_j(\mu)|\phi_i(\mu)> = <\phi_j(\nu)|K_i(\nu)|\phi_j(\nu)> \qquad (8.68)$$

The first–order variation in the functional J is

$$\delta J = 2\sum_i (<\delta\phi_i|h_\mu|\phi_i> + <\phi_i|h_\mu|\delta\phi_i>) \qquad (8.69)$$

$$+\sum_{i,j}(<\delta\phi_i|2J_j - K_j|\phi_i> + <\phi_i|2J_j - K_j|\delta\phi_i>)$$

$$+\sum_{i,j}(<\delta\phi_j|2J_i - K_i|\phi_j> + <\phi_j|2J_i - K_i|\delta\phi_j>)$$

$$-\sum_{i,j}(\lambda_{ij} <\delta\phi_i|\phi_j> + \lambda_{ij} <\phi_i|\delta\phi_j>) = 0$$

The first and second double summations are symmetric in their indices and lead to the same final sums. Thus

$$\delta J = 2\sum_i [<\delta\phi_i|h_\mu + \sum_j(2J_j - K_j)|\phi_i>] \qquad (8.70)$$

$$+2\sum_i [<\phi_i|h_\mu + \sum_j(2J_j - K_j)|\delta\phi_i>]$$

$$-\sum_{i,j}(\lambda_{ij} <\delta\phi_i|\phi_j> + \lambda_{ij} <\phi_i|\delta\phi_j>) = 0$$

Since h_μ , J_j , K_j are hermitian, the first and second summations are just the adjoints of each other; also $<\phi_j|\delta\phi_i>$ and $<\delta\phi_i|\phi_j>$ are adjoints of each other. Then Eq.(8.70) may be written as

$$\delta J = 2\sum_i [<\delta\phi_i|h_\mu + \sum_j(2J_j - K_j)|\phi_i> - \sum_j \lambda_{ij} <\delta\phi_i|\phi_j>] \quad (8.71)$$

$$+2\sum_i [<\delta\phi_i|h_\mu + \sum_j(2J_j - K_j)|\phi_i>^* - \sum_j \lambda_{ij} <\delta\phi_i|\phi_j>^*] = 0$$

This is satisfied by the conditions

$$[h_\mu + \sum_j(2J_j - K_j)]\phi_i = \sum_j \lambda_{ij}\phi_j \qquad (8.72)$$

$$[h_\mu + \sum_j(2J_j - K_j)]\phi_i^* = \sum_j \lambda_{ij}\phi_j^* \qquad (8.73)$$

These two equations, Eqs.(8.72) and (8.73), are complex conjugate of each other and equivalent. These equations are known as **Hartree–Fock equations**.

Eq.(8.72) may be written in matrix form as

$$\mathcal{F}\mathbf{\Phi} = \lambda\mathbf{\Phi} \tag{8.74}$$

where

$$\mathcal{F} = h_\mu + \sum_j (2J_j - K_j) \tag{8.75}$$

is called the **Hartree–Fock operator** (HF operator).

$$\mathbf{\Phi} = [\phi_1\phi_2\cdots\phi_N] \tag{8.76}$$

$$\lambda = \begin{bmatrix} \lambda_{11} & \cdots & \lambda_{1N} \\ \vdots & & \vdots \\ \lambda_{N1} & \cdots & \lambda_{NN} \end{bmatrix} \tag{8.77}$$

The eigenvalue equation for the HF operator F with the spatial orbitals $\{\phi_i\}$; $i = 1, 2, ..., N$

$$\mathcal{F}\phi_i = \varepsilon_i\phi_i \tag{8.78}$$

where ϕ_i are HF orbitals and ε_i are the corresponding HF eigenvalues.

$$\varepsilon_i = <\phi_i|\mathcal{F}|\phi_i> = \varepsilon_i^{(0)} + \sum_j (2J_{ij} - K_{ij}) \tag{8.79}$$

Total Hartree–Fock energy, E_{HF} is given by

$$E_{HF} = \sum_{i=1}^{N}(\varepsilon_i + \varepsilon_i^{(0)}) = 2\sum_{i=1}^{N}\varepsilon_i^{(0)} + \sum_{i,j}^{N}(2J_{ij} - K_{ij}) \tag{8.80}$$

One can write the total energy of a $(2N - 1)$–electron system as

$$E^+ = E_{HF}^{(k)}(2N - 1) = \sum_{i=1}^{N}(\varepsilon_i + \varepsilon_i^{(0)}) - \varepsilon_k \tag{8.81}$$

similar to this one, for a $(2N + 1)$–electron system

$$E^- = E_{HF}^{(m)}(2N + 1) = \sum_{i=1}^{N}(\varepsilon_i + \varepsilon_i^{(0)}) + \varepsilon_m \tag{8.82}$$

It then follows that

$$E^+ - E = -\varepsilon_k \tag{8.83}$$

$$E^- - E = +\varepsilon_m \tag{8.84}$$

where E represents the energy of the neutral system.

According to Koopman's theorem $-\varepsilon_k$ and $+\varepsilon_m$ can be regarded as an ionization potential, and the electron affinity, respectively, of the system described by $\psi(2N)$.

Hartree–SCF and Hartree–Fock methods give the He ground state energy as -2.86 $a.u.$ Although the variational method gives more accurate than this value, it is not practical for many–electron atoms; one has to use Hartree–SCF or Hartree–Fock methods.

The functions chosen to represent AOs ($\phi_{n\ell m_\ell}(r,\theta,\phi)$) are usually combinations of the spherical harmonics ($Y_{\ell m_\ell}(\theta,\phi)$) and one of the following radial functions ($R_{n\ell}(r)$):

1. Hydrogen–like functions (orthogonal)

$$R_{n\ell}(r) = r^\ell L_{n+\ell}^{2\ell+1}\left(\frac{2Zr}{n}\right)e^{-Zr/n} \quad ; \quad L \; : \; Laguerre \; functions \tag{8.85}$$

2. Slater functions (not orthogonal)

$$R_{n\ell}(r) = r^{n-1}e^{-\alpha r} \quad ; \quad n = 1, 2, ... \tag{8.86}$$

3. Gaussian functions (orthogonal)

$$R_{n\ell}(r) = r^{n+\ell}e^{-\alpha r^2} \quad ; \quad n = 0, 1, 2, ... \tag{8.87}$$

$$\phi_{n\ell m_\ell}(r,\theta,\phi) = R_{n\ell}(r)Y_{\ell m_\ell}(\theta,\phi) \tag{8.88}$$

8.5 The electron correlation energy

The HF energy is usually within about 1% of the experimental value (total energy). However, one is usually interested in energy differences, e.g., the energy difference between two spectroscopic states. Unfortunately, these energy differences themselves are often no larger than about 1% of the total energy of either state. Thus, small absolute errors in the total energies may easily lead to large relative errors in their differences. For this reason there is a tremendous amount of interest in quantum–mechanical calculations which give better energies than the HF method does. However, this method is used as a sort of reference point for more accurate calculations. Let us call E_{HF} the Hartree–Fock energy of a given state and ψ_{HF} the corresponding wave function. Both E_{HF} and ψ_{HF} are only approximations to the exact energy E_{exact} and the exact wave function ψ_{exact} of the non–relativistic Hamiltonian.

P.O. Löwdin (in 1959) defined the electron correlation energy as

$$E_{corr} = <H> (exact) - <H> (HF) \equiv E_{exact} - E_{HF} \qquad (8.89)$$

Correlation energy in HF method reflects the fact that Coulombic interaction between pairs of electrons, especially electrons with antiparallel spins, is not properly accounted for. Electrons of parallel spins are kept apart by the Pauli principle, an effect which overrides the Coulombic repulsion, and are thus described somewhat better than electrons of antiparallel spin.

The quantity E_{exact} is the exact energy of the non–relativistic Hamiltonian and hence is not quite the same as experimental energy. A certain amount of correlation is already included in the ψ_{HF} because of the fact that it is totaly antisymmetric. E_{corr} represents the correlation effects not included in the ψ_{HF}.

8.6   Many–electron theory of Sinanoğlu

It is well known that the HF method gives a fairly reasonable representation of the shell structures and electron densities of atoms. Relying partly on this observation, O. Sinanoğlu (in 1961) presented an analysis of the correlation problem by writing the exact wave function in the form

$$\psi = \phi_0 + \chi \qquad (8.90)$$

where ϕ_0 is the HF wave function and χ is a wave function representing the correlation correction. Furthermore,

$$< \phi_0 | \chi > = 0 \quad , \quad < \psi | \psi > = 1 + < \chi | \chi > \tag{8.91}$$

The HF energy in terms of spin orbitals can be written, in general,

$$E_{HF} = < \phi_0 | H | \phi_0 > = \sum_i^{2N} \varepsilon_i + \sum_{i<j}^{2N} (J_{ij} - K_{ij}) \tag{8.92}$$

using the wave function, Eq.(8.90), the exact energy can be written

$$E_{exact} = \frac{< \psi | H | \psi >}{< \psi | \psi >} \tag{8.93}$$

$$= E_{HF} + \frac{2 < \phi_0 | H - E_{HF} | \chi > + < \chi | H - E_{HF} | \chi >}{1 + < \chi | \chi >}$$

Sinanoğlu finds it convenient to introduce the following operators

$$\mathbf{e}_i = \mathcal{F} - \varepsilon_i \tag{8.94}$$

$$m_{ij} = \frac{1}{r_{ij}} + J_{ij} - K_{ij} - G_i(j) - G_j(i) \tag{8.95}$$

$$G_i(j) = J_i(j) - K_i(j) \tag{8.96}$$

The correlation energy is then given exactly by

$$E_{corr} = E_{exact} - E_{HF} \tag{8.97}$$

$$= \frac{2 < \phi_0 | \sum_{i<j} m_{ij} | \chi > + < \chi | \sum_i \mathbf{e}_i + \sum_{i<j} m_{ij} | \chi >}{1 + < \chi | \chi >}$$

The two–electron operator m_{ij} is called a **fluctuation potential**. The problem now is to find χ such that the total energy is minimized. The function χ can be represented exactly by an expansion in terms of functions which correspond to successively higher–excited configurations.

Sinanoğlu has shown that the first–order wave function $\psi^{(1)}$ of an N–electron system can be broken down into terms involving only pair functions. These pair functions satisfy equations just like those of an actual two–electron system expect that now each electron moves in the HF field of the entire system of particles.

The exact energy is approximated by

$$E \leq E_{HF} + \sum_{i<j} \varepsilon_{ij} \tag{8.98}$$

where ε_{ij} are pair correlation energies determined by separate minimizations for each pair of electrons. The advantage of the Sinanoğlu formulation is that it reduces the solution of one N–body problem to the solution of $\begin{pmatrix} N \\ 2 \end{pmatrix} = \frac{1}{2}N(N-1)$ two–body problems.

8.7 Slater Atomic Orbitals and screening constant rules

The simplest and the most practical method of calculating the energy levels of many–electron atoms is the Hartree–SCF procedure. In this method the atomic orbitals come out as numerical tables. Since such wave functions are not practical to use, J.C. Slater (in 1960) has proposed analytical functions,

$$\phi_{n\cdot\ell} = r^{n^*-1}e^{-Z_{n\ell}r/n^*}Y_{\ell m_\ell}(\theta,\phi) \tag{8.99}$$

which approximate the atomic orbitals, where $Y_{\ell m_\ell}(\theta,\phi)$ is a spherical harmonic in real form. These functions (STOs) are wave functions describing the motion of a single electron in a central field in which the potential energy is given by

$$V(r) = -\frac{Z_{n\ell}}{r} + \frac{n^*(n^*-1)}{2r^2} \tag{8.100}$$

The quantity n^* is an effective principal quantum number and $Z_{n\ell}$ is an effective nuclear charge. Both of these quantities are treated as empirical parameters. The Slater orbitals have no nodes in the radial portion whereas the hydrogen–like AOs have $n - \ell - 1$ such nodes.

Slater has shown that satisfactory values of n^* and $Z_{n\ell}$ may be chosen by use of some simple rules. For K, L, and M shells, one uses $n^* = n$. For N, O, and P shells n^* equals 3.7, 4.0, and 4.2, respectively.

The effective charge $Z_{n\ell}$ may be written

$$Z_{n\ell} = Z - S_{n\ell} \tag{8.101}$$

where Z is the atomic number and $S_{n\ell}$ is a screening constant depending upon the orbital $\phi_{n\ell}$ and the particular electronic configuration.

The screening constants are evaluated according to the following empirical rules:

1. Divide the orbitals into the following groups, each group having a different shielding constant:
   $(1s)$, $(2s, 2p)$, $(3s, 3p)$, $(3d)$, $(4s, 4p)$, $(4d)$, $(4f)$, $(5s, 5p)$

2. The shielding constant $S_{n\ell}$ is the sum of the following contributions:
   a) Zero from any shell outside the one considered.
   b) 0.35 from each other electron in the group considered, except that 0.3 is used for the $1s$ orbital.
   c) For s and p orbitals, 0.85 is subtracted for each electron in the next inner shell and 0.1 for all electrons still further in. If the orbital is d or f, 1.0 is subtracted for every electron within this orbital.

The Slater rules tend to become unreliable when applied to orbitals of principle quantum number 4 or greater. The above rules yield $Z_{1s} = 1.7$ for the ground state of He. For the K and L electrons of the ground state of the carbon atom one obtains 5.70 and 3.25, respectively.

$$S_{1s} = 0.3$$

$$S_{n\ell} = 0.35x + 0.85y + z \ ; \quad n > 1 \ , \quad \ell = 0, 1 \quad \text{(for } 2s, 2p, 3s, 3p)$$

$$x \equiv \# \text{ of remaining electrons in the same shell}$$

$y \equiv \#$ of electrons in the shell with p.q.n. $(n-1)$

$z \equiv \#$ of electrons with p.q.n. $\leq (n-2)$

$S_{3d} = 0.35x + y$

$x \equiv \#$ of remaining $3d$ electrons

$y \equiv \#$ of electrons with $n \leq 3$ and $\ell < 2$

Fore more accurate calculations the screening constants may be determined variationaly. The equations obtained by E. Clementi and D.L. Raimondi (in 1963) are (valid for $Z = 2 - 36$ and ground state configurations):

$$S_{1s} = 0.3[N(1s) - 1] + 0.0072[N(2s) + N(2p)] \tag{8.102}$$

$$+0.0158[N(3s) + N(3p) + N(4s) + N(3d) + N(4p)]$$

$$S_{2s} = 1.7208 + 0.3601[N(2s) + N(2p) - 1] \tag{8.103}$$

$$+0.2062[N(3s) + N(3p) + N(4s) + N(3d) + N(4p)]$$

$$S_{2p} = 2.5787 + 0.3326[N(2p) - 1] - 0.0773N(3s) \tag{8.104}$$

$$-0.0161[N(3p) + N(4s)] - 0.0048N(3d) + 0.0085N(4p)$$

$$S_{3s} = 8.4927 + 0.2501[N(3s) + N(3p) - 1] + 0.0778N(4s) \tag{8.105}$$

$$+0.3382N(3d) + 0.1978N(4p)$$

$$S_{3p} = 9.3345 + 0.3803[N(3p) - 1] + 0.0526N(4s) \tag{8.106}$$

$$+0.3289N(3d) + 0.1558N(4p)$$

$$S_{4s} = 15.1505 + 0.0971[N(4s) - 1] \tag{8.107}$$

$$+0.8433N(3d) + 0.0687N(4p)$$

$$S_{3d} = 13.5894 + 0.2693[N(3d) - 1] - 0.1065N(4p) \tag{8.108}$$

$$S_{4s} = 24.7782 + 0.2905[N(4p) - 1] \tag{8.109}$$

Table 8.1: Values of $Z_{n\ell}$ in Slater AOs:

Z	Atom	$Z_{n\ell}$		
		$1s$	$(2s, 2p)$	$(3s, 3p)$
1	H	1.00		
2	He	1.70		
3	Li	2.70	1.30	
4	Be	3.70	1.95	
5	B	4.70	2.60	
6	C	5.70	3.25	
7	N	6.70	3.90	
8	O	7.70	4.55	
9	F	8.70	5.20	
10	Ne	9.70	5.85	
11	Na	10.70	6.85	2.20
12	Mg	11.70	7.85	2.85
13	Al	12.70	8.85	3.50
14	Si	13.70	9.85	4.15
15	P	14.70	10.85	4.80

8.8   Roothaan method; HF method for molecules

The basic approach to the calculation of molecular orbitals consists of efforts to solve the Hartree–Fock equations which are formally identical for atoms and molecules. We have seen that the problem is already quite complicated for atoms; in molecules, the absence of three–dimensional rotational symmetry and the occurrence of multicenter integrals increase the complexity to an extent that direct, iterative, solutions of the HF equations are almost impossible. Instead, one postulates that molecular orbitals are to be expressed as linear combinations of atomic orbitals (LCAO). This is known as the Roothaan method. We will consider the closed shell molecules, actually many molecules in their ground state have close shell configurations. Since all the electrons in a closed shell have their spins paired ($S = 0$) the total antisymmetric wave function can be represented by a single Slater determinant

$$
\Psi = \frac{1}{\sqrt{(2N)!}}
\begin{vmatrix}
\psi_1(1) & \bar{\psi}_1(1) & \cdots & \psi_N(1) & \bar{\psi}_N(1) \\
\psi_1(2) & \bar{\psi}_1(2) & \cdots & \psi_N(2) & \bar{\psi}_N(2) \\
\vdots & \vdots & & \vdots & \vdots \\
\psi_1(2N) & \bar{\psi}_1(2N) & \cdots & \psi_N(2N) & \bar{\psi}_N(2N)
\end{vmatrix}
\tag{8.110}
$$

As in the atomic case, the objective is to find the orbitals that will minimize the energy

$$
E = < \Psi|H|\Psi >
\tag{8.111}
$$

where H is the complete non–relativistic electronic Hamiltonian of the molecule.

The molecular orbitals that satisfy the variational principle are obtained from solutions of the Hartree–Fock equations; for a closed shell they are of the form

$$
F(1)\psi_i(1) = \varepsilon_i \psi_i(1)
\tag{8.112}
$$

where

$$
F(1) = h_0(1) + \sum_k [2J_k(1) - K_k(1)]
\tag{8.113}
$$

$$
h_0 = -\frac{\hbar^2}{2m}\nabla_1^2 - \sum_\mu \frac{Z_\mu e^2}{r_{1_\mu}}
\tag{8.114}
$$

J_k and K_k are the Coulomb and exchange operators, respectively, which depend on the molecular orbitals themselves and are defined as

$$
J_k(1)\psi_i(1) = < \psi_k(2)|\frac{e^2}{r_{12}}|\psi_k(2) > \psi_i(1)
\tag{8.115}
$$

$$
K_k(1)\psi_i(1) = < \psi_k(2)|\frac{e^2}{r_{12}}|\psi_i(2) > \psi_k(1)
\tag{8.116}
$$

The total energy is

$$
E = 2\sum_i \varepsilon_i^{(0)} + \sum_{i,k}[2J_{ik} - K_{ik}]
\tag{8.117}
$$

We now introduce the constraint, due to Roothaan (1951), that the molecular orbitals are to be expressed as linear combinations of atomic orbitals:

$$\psi_i = \sum_\mu C_{\mu i}\phi_\mu \tag{8.118}$$

or in matrix form

$$\boldsymbol{\Psi} = \boldsymbol{\Phi}\mathbf{C} \tag{8.119}$$

where $\boldsymbol{\Psi}$ and $\boldsymbol{\Phi}$ are now matrices and $\mathbf{C}$ is a square matrix. The ϕ_μ are the atomic orbitals and they are centered on various nuclei. The ϕ_μ are not necessarily orthogonal. Hence we define

$$S_{\mu\nu} =< \phi_\mu|\phi_\nu > \tag{8.120}$$

as the overlap integral for the orbitals ϕ_μ and ϕ_ν . The molecular orbitals, on the other hand, are required to satisfy the orthogonality condition

$$< \psi_i|\psi_j >= \delta_{ij} \tag{8.121}$$

so that in the LCAO approximation,

$$\sum_{\mu,\nu} C_{\mu i}^* C_{\nu j} < \phi_\mu|\phi_\nu >= \sum_{\mu,\nu} C_{\mu i}^* C_{\nu j} S_{\mu\nu} = \delta_{ij} \tag{8.122}$$

which in matrix form

$$\mathbf{C}^+\mathbf{SC} = \mathbf{1} \tag{8.123}$$

The charge density at the point $\mathbf{r}$ is defined by

$$\rho(\mathbf{r}) = 2\sum_i^{occ} \psi_i^*(\mathbf{r})\psi_i(\mathbf{r}) \tag{8.124}$$

the factor of 2 appearing because of the double occupancy of the molecular orbitals by two electrons with opposite spin. In terms of atomic orbitals

$$\rho(\mathbf{r}) = 2\sum_i \sum_{\mu,\nu} C_{\mu i}^* C_{\nu i}\phi_\mu^*(\mathbf{r})\phi_\nu(\mathbf{r}) = \sum_{\mu,\nu} P_{\mu\nu}\phi_\mu^*(\mathbf{r})\phi_\nu(\mathbf{r}) \tag{8.125}$$

where

$$P_{\mu\nu} = 2\sum_i^{occ} C_{\mu i}^* C_{\nu i} \tag{8.126}$$

is the density matrix.

If N is the number of occupied (spatial) orbitals, the closed shell will contain $2N$ electrons; therefore the integral of the charge density over all space must be equal to $2N$:

$$\int \rho(\mathbf{r})d\mathbf{r} = \sum_{\mu,\nu} P_{\mu\nu} \int \phi_\mu^*(\mathbf{r})\phi_\nu(\mathbf{r})d\mathbf{r} = \sum_{\mu,\nu} P_{\mu\nu}S_{\mu\nu} = 2N \qquad (8.127)$$

With these relations

$$h_0(1)\psi_i(1) = \sum_\nu h_0(1)C_{\nu i}\phi_\nu(1) \qquad (8.128)$$

$$J_k(1)\psi_i(1) = \sum_{\lambda\sigma\nu} C_{\lambda k}^* C_{\sigma k}C_{\nu i} < \phi_\lambda(2)|\frac{e^2}{r_{12}}|\phi_\sigma(2) > \phi_\nu(1) \qquad (8.129)$$

$$K_k(1)\psi_i(1) = \sum_{\lambda\sigma\nu} C_{\lambda k}^* C_{\sigma k}C_{\nu i} < \phi_\lambda(2)|\frac{e^2}{r_{12}}|\phi_\nu(2) > \phi_\sigma(1) \qquad (8.130)$$

$$F(1)\psi_i(1) = \sum_\nu h_0(1)C_{\nu i}\phi_\nu(1) + \sum_{k\lambda\sigma\nu} C_{\lambda k}^* C_{\sigma k}C_{\nu i} \qquad (8.131)$$

$$\times [2 < \phi_\lambda(2)|\frac{e^2}{r_{12}}|\phi_\sigma(2) > \phi_\nu(1) - < \phi_\lambda(2)|\frac{e^2}{r_{12}}|\phi_\nu(2) > \phi_\sigma(1)]$$

$$= \sum_\nu \varepsilon_i C_{\nu i}\phi_\nu(1)$$

This equation is the HF equation in LCAO form. Let us introduce the notation

$$< \phi_\mu(1)\phi_\lambda(2)|\frac{e^2}{r_{12}}|\phi_\nu(1)\phi_\sigma(2) > \equiv < \mu\nu|\lambda\sigma > \qquad (8.132)$$

This definition implies

$$< \mu\nu|\lambda\sigma > = < \lambda\sigma|\mu\nu > \quad , \quad < \mu\nu|\lambda\sigma >^* = < \nu\mu|\sigma\lambda >^* \qquad (8.133)$$

Multiplying both sides of Eq.(8.131) by $\phi_\mu^*(1)$ and integrating over the coordinates of electron (1) we obtain

$$\sum_\nu C_{\nu i}H_{\mu\nu} + \sum_{k\lambda\sigma\nu} C_{\lambda k}^* C_{\sigma k}C_{\nu i}[2 < \mu\nu|\lambda\sigma > - < \mu\sigma|\lambda\nu >] \qquad (8.134)$$

$$= \sum_{\nu} \varepsilon_i C_{\nu i} S_{\mu\nu}$$

in which

$$H_{\mu\nu} = <\phi_\mu(1)|h_0(1)|\phi_\nu(1)> \tag{8.135}$$

inserting the density matrix,

$$\sum_{\nu} C_{\nu i} H_{\mu\nu} + \sum_{\lambda\sigma\nu} P_{\lambda\sigma}[<\mu\nu|\lambda\sigma> -\frac{1}{2}<\mu\sigma|\lambda\nu>] = \sum_{\nu} \varepsilon_i C_{\nu i} S_{\mu\nu} \tag{8.136}$$

Finally, by writing

$$F_{\mu\nu} = H_{\mu\nu} + \sum_{\lambda\sigma} P_{\lambda\sigma}[<\mu\nu|\lambda\sigma> -\frac{1}{2}<\mu\sigma|\lambda\nu>] \tag{8.137}$$

the HF equations assume the compact form

$$\sum_{\nu} (F_{\mu\nu} - \varepsilon_i S_{\mu\nu}) C_{\nu i} = 0 \tag{8.138}$$

or in matrix form,

$$(\mathbf{F} - \varepsilon\mathbf{S})\mathbf{C} = 0 \tag{8.139}$$

where ε is a diagonal matrix whose diagonal elements ε_i are the orbital energies. The set of homogeneous equations will have nontrivial solutions only when

$$|F_{\mu\nu} - \varepsilon_i S_{\mu\nu}| = 0 \tag{8.140}$$

The LCAO approximation has converted the original partial differential equations (Hartree–Fock) into algebraic equations (Roothaan).

The total energy of a closed shell was shown to be

$$E = 2\sum_{i} I_i + \sum_{i,k}[2J_{ik} - K_{ik}] \tag{8.141}$$

In the LCAO appproximation we have

$$\varepsilon_i^{(0)} = <\psi_i(1)|h_0(1)|\psi_i(1)> = \sum_{\mu\nu} C_{\mu i}^* C_{\nu i} H_{\mu\nu} \tag{8.142}$$

$$J_{ik} = <\psi_i(1)\psi_k(2)|\frac{e^2}{r_{12}}|\psi_i(1)\psi_k(2)> \tag{8.143}$$

$$= \sum_{\mu\lambda\nu\sigma} C^*_{\mu i} C^*_{\lambda k} C_{\nu i} C_{\sigma k} < \mu\nu|\lambda\sigma >$$

$$K_{ik} =< \psi_i(1)\psi_k(2)| \frac{e^2}{r_{12}} |\psi_k(1)\psi_i(2) > \qquad (8.144)$$

$$= \sum_{\mu\lambda\nu\sigma} C^*_{\mu i} C^*_{\lambda k} C_{\nu i} C_{\sigma k} < \mu\sigma|\lambda\nu >$$

Substituting in total energy, E, the expression for the energy becomes

$$E = \sum_{\mu\nu} P_{\mu\nu} H_{\mu\nu} + \frac{1}{2} \sum_{\mu\nu\lambda\sigma} P_{\mu\nu} P_{\lambda\sigma} [< \mu\nu|\lambda\sigma > -\frac{1}{2} < \mu\sigma|\lambda\nu >] \quad (8.145)$$

$$= \frac{1}{2} \sum_{\mu\nu} P_{\mu\nu} (H_{\mu\nu} + F_{\mu\nu})$$

8.9   Computational methods for molecules

The Roothaan equations into which the HF equations have evolved as a result of the LCAO approximation still require an iterative procedure because the matrix elements $F_{\mu\nu}$ depend on unknown coefficients (or density matrices). A typical calculation involves the following steps:

1. Specify a set of atomic orbitals (the basis set). Among the popular choices are the Slater orbitals (STO) and Gaussian orbitals.

2. Compute the overlap integrals $S_{\mu\nu}$.

3. Compute the core Hamiltonian matrix $H_{\mu\nu}$.

4. Compute the two–electron integrals $< \mu\nu|\lambda\sigma >$.

5. Compute the eigenvectors of $H_{\mu\nu}$. This gives a starting set of LCAO coefficients (and the initial set of molecular orbitals).

6. Assign electrons in pairs to the lowest molecular orbitals until all the electrons have been assigned.

7. Compute the density matrix $P_{\mu\nu}$.

8. Compute the total electronic energy

$$E = \sum_{\mu\nu} P_{\mu\nu} H_{\mu\nu} + \frac{1}{2} \sum_{\mu\nu\lambda\sigma} P_{\mu\nu} P_{\lambda\sigma} [< \mu\nu|\lambda\sigma > - \frac{1}{2} < \mu\lambda|\nu\sigma >]$$

(8.146)

9. Compute the Fock Hamiltonian

$$F_{\mu\nu} = H_{\mu\nu} + \sum_{\lambda\sigma} P_{\lambda\sigma} [< \mu\nu|\lambda\sigma > - \frac{1}{2} < \mu\lambda|\nu\sigma >] \qquad (8.147)$$

10. Compute the eigenvectors of $F_{\mu\nu}$ to obtain a second set of LCAO coefficients.

11. Continue until the total energy E remains constant to the required accuracy.

Despite the simplification in the HF formalism resulting from the introduction of the LCAO approximation, the most series limitation in MO calculations is the large number of integrals that must be evaluated. Many integrals are of the two–electron, multicenter type which consume most of the computer time. There are some approximations which reduce the number of integrals without lossing the accuracy of the calculations. We will summarize some of these approaches here.

1- Neglect of differential overlap (NDO). This consists of setting $\phi_\mu(1)\phi_\nu(1) = 0$ for $\mu \neq \nu$ in the electron repulsion integrals, and has the effect of eliminating all three– and four–center integrals.

2- Neglect of all but valence electrons. Atomic orbitals, chosen exclusively from the valence shell of each atom, constitute the basis set for the LCAO expansion. Inner shell electrons, because they are more tightly bound, do not participate to any significant extent in the formation of molecular orbitals. Empirically, it is known that chemical effects are largely due to the valence electrons.

3- Except in the very simplest cases, some use is made of experimental information such as atomic ionization potentials and electron affinities. These are used to assign numerical values to a number of theoretical parameters. A popular version of MO theory is known as the

CNDO method (complete neglect of differential overlap) in which condition $\phi_\mu(1)\phi_\nu(1) = 0$, $(\mu \neq \nu)$ is used for all two–electron interaction integrals. In this scheme, using the convention that ϕ_μ is on atom A and ϕ_ν on atom B,

$$F_{\mu\mu} = U_{\mu\mu} + \frac{1}{2}P_{\mu\mu}\gamma_{\mu\mu} + \sum_{\nu \neq \mu}(P_{\nu\nu} - Z_\nu)\gamma_{\mu\nu} \tag{8.148}$$

$$F_{\mu\nu} = H_{\mu\nu} - \frac{1}{2}P_{\mu\nu}\gamma_{\mu\nu} \quad (\mu \neq \nu) \tag{8.149}$$

In these expressions $U_{\mu\mu}$ is a core integral which is approximated by $U_{\mu\mu} = -I_\mu$, where I_μ is the atomic ionization potential for the orbital ϕ_μ. The condition of zero differential overlap is employed so that

$$< \mu\lambda|\nu\sigma >=< \mu\mu|\nu\nu > \delta_{\mu\lambda}\delta_{\nu\sigma} \tag{8.150}$$

$\gamma_{\mu\nu}$ is the two–electron repulsion integral, $\gamma_{\mu\nu} =< \mu\mu|\nu\nu >$. When $\mu = \nu$, the integrals are approximated by

$$\gamma_{\mu\mu} = I_\mu - A_\mu \tag{8.151}$$

where A_μ is the electron affinity in the orbital ϕ_μ. $H_{\mu\nu}$ is a core resonance integral which is often approximated by setting

$H_{\mu\nu} = 0$ for μ and ν not nearest neighbors,

$H_{\mu\nu} = $ empirical constant when μ and ν are bonded nearest neighbors.

Z_ν is the core charge of atom B and is equal to the nuclear charge minus the number of inner shell electrons.

Another approximation which is worth to mention is the Extended Hückel Method (EHM) in which all, or almost all, integrals are replaced by empirical parameters thereby eliminating the self–consistent, iterative procedure. If the secular equation $|F_{\mu\nu} - \varepsilon_i S_{\mu\nu}| = 0$ is written as $|H_{\mu\nu} - \varepsilon_i S_{\mu\nu}| = 0$, the diagonal elements are assigned empirical values depending on the orbitals and the type of molecule. The off–diagonal elements are approximated by

$$H_{\mu\nu} = \frac{1}{2}K(H_{\mu\mu} + H_{\nu\nu})S_{\mu\nu} \tag{8.152}$$

in which K is an empirical parameter and $S_{\mu\nu}$ the overlap integral for atomic orbitals ϕ_μ and ϕ_ν centered on the appropriate atoms. Having made the numerical assignments; solutions to $|H_{\mu\nu} - \varepsilon_i S_{\mu\nu}| = 0$ yield the eigenvalues ε_i and hence the coefficients $C_{\mu i}$ in the molecular orbitals $\psi_i = \sum_\mu C_{\mu i} \phi_\mu$.

8.10   Worked examples

Example - 8.1 :

Using Slater rules calculate (a) the ionization potentials of lithium atom, (b) the $2p$ orbital radius of carbon atom to estimate the atomic size.

Solution :

a) $Li(1s^2 2s)$: $Z_{1s} = 2.7$, $Z_{2s} = 1.3$; $E_{Li} = 2E_{1s} + E_{2s}$

$E_{Li} = [-2(2.7/1)^2 - (1.3/2)^2] \times 13.6 \; eV = -204.034 \; eV$

$Li^+(1s^2)$: $Z_{1s} = 2.7$; $E_{Li^+} = 2E_{1s}$

$E_{Li^+} = -2(2.7/1)^2 \times 13.6 \; eV = -198.288 \; eV$

$Li^{++}(1s)$: $Z_{1s} = 3.0$; $E_{Li^{++}} = -(3.0/1)^2 \times 13.6 \; eV = -122.4 \; eV$

$IP_1 = E_{Li^+} - E_{Li} = 5.746 \; eV$; $(IP_1)_{exp} = 5.39 \; eV$

$IP_2 = E_{Li^{++}} - E_{Li^+} = 75.888 \; eV$; $(IP_2)_{exp} = 75.7 \; eV$

$IP_3 = -E_{Li^{++}} = 122.4 \; eV$; $(IP_3)_{exp} = 122.4 \; eV$

b) $C(1s^2 2s^2 2p^2)$, $Z_{2p} = 3.25$

$$r_{eff} = \frac{n^2}{Z_{nl}} a_0 = \frac{4}{3.25} a_0 = 1.23 a_0 = 0.65 \; \mathring{A}$$

$(r_{eff})_{exp} \cong 0.62 \ \mathring{A}$

It is a good approximation.

Example - 8.2 :

Calculate the ionization potentials of Li atom using the screening constants determined by Clementi and Raimondi.

Solution :

$Li(1s^2 2s)$:
$S_{1s} = 0.3 \times (2 - 1) + 0.0072 \times (1) = 0.3072$
$Z_{1s} = Z - S_{1s} = 3 - 0.3072 = 2.6928$
$S_{2s} = 1.7208 + 0.3601 \times (1 - 1) = 1.7208$
$Z_{2s} = Z - S_{2s} = 3 - 1.7208 = 1.2792$
$E_{Li} = 2E_{1s} + E_{2s} = [-2 \times (2.6928/1)^2 - (1.2792/2)^2] \times 13.6 \ eV = -202.795 \ eV$

$Li^+(1s^2)$:
$S_{1s} = 0.3 \times (2 - 1) = 0.3$
$Z_{1s} = Z - S_{1s} = 3 - 0.3 = 2.7$
$E_{Li^+} = 2E_{1s} = -2 \times (2.7/1)^2 \times 13.6 \ eV = -198.288 \ eV$

$Li^{++}(1s)$:
$S_{1s} = 0.3 \times (1 - 1) = 0$
$Z_{1s} = Z - S_{1s} = 3.0$
$E_{Li^{++}} = E_{1s} = -(3.0/1)^2 \times 13.6 \ eV = -122.4 \ eV$

$IP_1 = E_{Li^+} - E_{Li} = 4.507 \ eV$; $(IP_1)_{exp} = 5.39 \ eV$

$IP_2 = E_{Li^{++}} - E_{Li^+} = 75.888 \ eV$; $(IP_2)_{exp} = 75.7 \ eV$

$IP_3 = -E_{Li^{++}} = 122.4 \ eV$; $(IP_3)_{exp} = 122.4 \ eV$

Example - 8.3 :

Using Slater rules calculate (a) the screening constants, (b) the first ionozation potential, and (c) the atomic size for silicon atom.

Solution :

a) Si configuration: $1s^2 2s^2 2p^6 3s^2 3p^2$, $Z = 14$

$S_{1s} = 0.3$, $Z_{1s} = Z - S_{1s} = 14 - 0.3 = 13.7$
$S_{2s} = S_{2p} = 0.35x + 0.85y + z = 0.35 \times 7 + 0.85 \times 2 + 0 = 4.15$
$Z_{2s} = Z_{2p} = Z - S_{2s} = 14 - 4.15 = 9.85$
$S_{3s} = S_{3p} = 0.35x + 0.85y + z = 0.35 \times 3 + 0.85 \times 8 + 2 = 9.85$
$Z_{3s} = Z_{3p} = Z - S_{3s} = 14 - 9.85 = 4.15$

b) Si^+ configuration: $1s^2 2s^2 2p^6 3s^2 3p$, $Z = 14$

$S_{1s}(Si^+) = S_{1s}(Si)$, $S_{2s}(Si^+) = S_{2s}(Si)$, $S_{2p}(Si^+) = S_{2p}(Si)$

$S_{3s}(Si^+) \neq S_{3s}(Si)$, $S_{3p}(Si^+) \neq S_{3p}(Si)$

$S_{3s} = S_{3p} = 0.35x + 0.85y + z = 0.35 \times 2 + 0.85 \times 8 + 2 = 9.54$
$Z_{3s} = Z_{3p} = Z - S_{3s} = 14 - 9.54 = 4.46$
$Z_{1s} = 13.7$, $Z_{2s} = Z_{2p} = 9.85$

$$IP_1 = (2E_{1s} + 8E_{2s} + 3E_{3s})_{Si^+} - (2E_{1s} + 8E_{2s} + 4E_{3s})_{Si}$$

$$= 3E_{3s}(Si^+) - 4E_{3s}(Si)$$

$$= 3\left(-\frac{Z_e^2}{2 \times 3^2}\right)_{Si^+} - 4\left(-\frac{Z_e^2}{2 \times 3^2}\right)_{Si} = -\frac{3 \times (4.46)^2}{2 \times 3^2} + \frac{4 \times (4.15)^2}{2 \times 3^2} \ a.u.$$

$$\cong -3.3153 + 3.8272 = 0.5119 \ a.u. = 0.5119 \times 27.21 \ eV \cong 13.93 \ eV$$

$(IP_1)_{exp} = 8.149 \ eV$

c)
$$r_{eff} = \frac{n^2}{Z_e} a_0 = \frac{e^2}{4.15} a_0 = \frac{e^2}{4.15} \times 0.529 \ \mathring{A} \cong 1.147 \ \mathring{A}$$

$r_{orb} \cong 1.068 \ \mathring{A}$ (quantum mechanical calculation).

Example - 8.4 :

Consider the ground state of a neutral atom with nuclear charge Z. Assume that all subshells with the same p.q.n. n are filled; that is, we are considering closed shell atoms like He, Ne, Ar, etc. Further, assume that the electrons are non–interacting. The orbital energies of the atoms then become hydrogenic.

a) Show that the forgoing assumptions lead to the relation $Z = n_0(n_0+1)(2n_0+1)/3$, where n_0 is the p.q.n. of the highest occupied atomic orbital.

b) Determine the ground state energy in terms of n_0.

c) The Hartree–Fock ground state energies for He and Ne are -2.86 and -128.5 a.u., respectively. Calculate the ground state energy for He and Ne, using the energy expression from (b). Comment on the discrepancy between those energies and the HF energies.

d) The effect of electronic screening is often approximately taken into account by using screened hydrogenic orbitals rather than exact hydrogenic orbitals. In a screened hydrogenic orbital the orbital exponent is $Z_{n\ell} = Z - S_{n\ell}$, where $S_{n\ell}$ is the screening constant. Use Slater's screening constants to calculate the ground state energies of He and Ne. Comment on the accuracy of this approach relative to HF and the non–screened approach.

Solution :

a) The number of electrons is

$$Z = 2\sum_{n=1}^{n_0}\sum_{\ell=0}^{n-1}\sum_{m_\ell=-\ell}^{\ell} 1 = 2\sum_{n=1}^{n_0}\sum_{\ell=0}^{n-1}(2\ell+1) = 2\sum_{n=1}^{n_0} n^2 = n_0(n_0+1)\frac{(2n_0+1)}{3}$$

b) The number of electrons with p.q.n. n is $N_n = 2n^2$. The orbital energy of such an electron is $\varepsilon_n = -Z^2/(2n^2)$. The ground state energy becomes

$$E = \sum_{n=1}^{n_0} \varepsilon_n N_n = -n_0 Z^2 = -n_0^3(n_0+1)^2\frac{(2n_0+1)^2}{9}$$

c) $E_{He} = -4$ a.u. , $n_0 = 1$, $E_{HF} = -2.86$ a.u.

$E_{Ne} = -200$ a.u. , $n_0 = 2$, $E_{HF} = -128.5$ a.u.

Both the HF and the present hydrogenic approximation are independent–particle models. In the former, an electron moves in the average potential from all the other electrons. This average $e - e$ interaction, which is absent in the hydrogenic approximation, is thus very important for obtaining a realistic independent–particle model description.

d) $Z_{1s}(He) = 1.70 \quad \rightarrow \quad E_{He} = -2.89$ a.u.

$Z_{1s}(Ne) = 9.7$, $Z_{2s}(Ne) = Z_{2p}(Ne) = 5.85 \quad \rightarrow \quad E_{Ne} = -128.31$ a.u.

The screened results are in very close agreement with the HF energies.

Example - 8.5 :

In semiempirical molecular orbital methods, such as Hückel and extended Hückel methods, the ground–state total energy is usually estimated as $E_0 = \sum_r n_r \varepsilon_r$, where n_r is the occupation number of molecular spin orbital ψ_r; that is, $n_r = 1$ if ψ_r is an occupied molecular orbital and $n_r = 0$ if ψ_r is an unoccupied molecular spin orbital. ε_r is the molecular orbital energy. Let V_{nn} and V_{ee} denote the average value of the nuclear and electron repolsion operator with respect to the exact Hartree–Fock ground state. If we assume that the Hartree–Fock total energy can be approximated as above, where ε denotes the Hartree–Fock orbital energies, show then that this assumption implies that $V_{nn} \cong V_{ee}$.

Solution :

The exact Hartree–Fock total energy can be written as

$$E_{HF} = 2\sum_{k}^{occ} \varepsilon_k - \sum_{kl}^{occ}[2(kk|ll) - (kl|lk)] + \sum_{\mu>\nu} \frac{Z_\mu Z_\nu}{R_{\mu\nu}}$$

or $E_{HF} = \sum_r n_r \varepsilon_r + V_{nn} - V_{ee}$

and $V_{nn} \cong V_{ee}$ follows straightforward when $E_0 = E_{HF}$

Example - 8.6 :

Assuming that the Born–Oppenheimer approximation is valid, determine the conditions under which the potential energy curves for two different electronic states of a diatomic molecule can cross. (Hint: Assume that all of the wave functions are known except those associated with the two states in question. Then expand these two states in terms of the functions ψ_1 and ψ_2 which are orthogonal to each other and to all of the known wave functions.)

Solution :

As suggested we expand the unknown states in term of ψ_1 and ψ_2: $u = c_1\psi_1 + c_2\psi_2$. The coefficients can be determined by solving the 2×2 secular determinant:

$$\begin{vmatrix} H_{11} - E & H_{12} \\ H_{12} & H_{22} - E \end{vmatrix} = 0$$

The solutions are $u = \frac{1}{\sqrt{2}}(\psi_1 \pm \psi_2)$ and

$$E_{\pm} = \frac{1}{2}(H_{11} + H_{22}) \pm \frac{1}{2}\sqrt{(H_{22} - H_{11})^2 + 4H_{12}^2}$$

Assuming that $[2H_{12}/(H_{22} - H_{11})]^2 < 1$ and taking $H_{22} > H_{11}$, $E_{\pm}$ can be expanded to obtain:

$$E_1 = H_{11} - \frac{H_{12}^2}{H_{22} - H_{11}} + \cdots \quad , \quad E_2 = H_{22} + \frac{H_{12}^2}{H_{22} - H_{11}} - \cdots$$

In order for the levels to cross for some value of R so that $E_1 = E_2$ we must have $H_{11} - H_{22} = 0$ and $H_{12} = 0$. Otherwise there will be no simultaneous solutions for those equations. If ψ_1 and ψ_2 have different symmetry properties, $H_{12} = 0$ since H is totaly symmetric. In general, if ψ_1 and ψ_2 have the same symmetry properties $H_{12} \neq 0$ and no crossing can occur. States having different multiplicities have different symmetry properties and thus may cross.

Bibliography and References

Textbooks:

There are many textbooks about atomic and molecular physics. We have listed some selected books in this field. The list given here (in chronological order) is not meant to be exhaustive, and no book is criticized by its omission.

1. H.E. White, *Introduction to Atomic Spectra*, McGraw-Hill, 1934.

2. L. Pauling and E.B. Wilson, Jr., *Introduction to Quantum Mechanics*, McGraw–Hill, 1935.

3. G. Herzberg, *Atomic Spectra and Atomic Structure*, Dover, 1944.

4. G. Herzberg, *Spectra of Diatomic Molecules*, Van Nostrand, 1950.

5. D.R. Hartree, *The Calculation of atomic Structures*, Wiley, 1957.

6. L. Pauling, *The Nature of the Chemical Bond*, Cornell University Press, 1960.

7. J.C. Slater, *Quantum Theory of Atomic Structure*, McGraw-Hill, 1960.

8. R.M. Eisberg, *Fundamentals of Modern Physics*, Wiley, 1961.

9. H.G. Kuhn, *Atomic Spectra*, Academic Press, 1962.

10. J.C. Slater, *Quantum Theory of Molecular Structure*, McGraw-Hill, 1963.

11. F.L.Pilar, *Elementary Quantum Chemistry*, McGraw-Hill, 1968.

12. M. Karplus and R.N. Porter, *Atoms and Molecules*, Benjamin, 1970.

13. U. Fano and L. Fano, *Physics of Atoms and Molecules*, The University of Chicago Press, 1972.

14. M.A. Morrison, T.L. Estle, and N.F. Lane, *Quantum States of Atoms, Molecules, and Solids*, Prentice-Hall, 1976.

15. H.A. Bethe and E.E. Salpeter, *Quantum Mechanics of One- and Two-Electron Atoms*, Plenum, 1977.

16. C.F. Fischer, *The Hartree–Fock Method for Atoms*, Wiley, 1977.

17. M. Weissbluth, *Atoms and Molecules*, Academic Press, 1978.

18. I.I. Sobelman, *Atomic Spectra and Radiative Transitions*, Springer-ger Verlag, 1979.

19. E.U. Condon and H. Odabaşı, *Atomic Structure*, Cambridge University Press, 1980.

20. B.H. Bransden and C.J. Joachain, *Physics of Atoms and Molecules,les* Longman, 1983.

21. P.W. Atkins, *Molecular Quantum Mechanics*, Oxford University Press, 1983.

22. H. Friedrich, *Theoretical Atomic Physics*, Springer–Verlag, 1990.   ·

The following selected books are related to various subjects in the present   t book (in chronological order).

1. E.U. Condon and G.H. Shortley, *The Theory of Atomic Spectra*, Cambridge University Press, 1935.

2. A.R. Edmonds, *Angular Momentum in Quantum Mechanics*, Princeton University Press, 1957.

3. M.E. Rose, *Elementary Theory of Angular Momentum*, Wiley, 1957.

4. P.A.M. Dirac, *The Principles of Quantum Mechanics*, Oxford University Press, 1958.

5. E.P. Wigner, *Group Theory and its Applications to the Quantum Mechanics of Atomic Spectra*, Academic Press, 1959.

6. C.A. Coulson, *Valence*, Oxford University Press, 1961.

7. R.G. Parr, *The Quantum Theory of Molecular Structure*, Benjamin, 1963.

8. H.A. Bethe and R.W. Jackiw, *Intermediate Quantum Mechanics*, Benjamin, 1968.

9. G. Baym, *Lectures on Quantum Mechanics*, Benjamin, 1969.

10. R. McWeeny and B.T. Sutcliffe, *Methods of Molecular Quantum Mechanics*, Academic, 1969.

11. M. Mizushima, *Quantum Mechanics of Atomic Spectra and Atomic Structure*, Benjamin, 1970.

12. J.N. Murrell and A.J. Harget, *Semi–empirical Self–consistent–field Molecular–orbital Theory of Molecules*, Wiley, 1972.

13. B.R. Judd, *Angular Momentum Theory for Diatomic Molecules*, Academic, 1975.

14. N.H. March, *Self–Consistent Fields in Atoms*, Benjamin, 1975.

15. J.D. Jackson, *Classical Electrodynamics*, Wiley, 1975.

16. C. Cohen–Tannoudji, B. Diu and F. Laloe, *Quantum Mechanics*, Wiley, 1977.

17. D. Papousek and M.R. Aliev, *Molecular Vibrational–Rotational Spectra*, Elsevier, 1982.

18. W. Heitler, *The Quantum Theory of Radiation*, Dover, 1984.

19. J.I. Steinfeld, *Molecules and Radiation*, MIT Press, 1985.

20. H. Lefebvre–Brion and R.W. Field, *Perturbations in the Spectra of Diatomic Molecules*, Academic, 1986.

21. M. Weissbluth, *Photon–Atom Interactions*, Academic, 1989.

22. L. Szasz, *The Electronic Structure of Atoms*, Wiley, 1992.

23. C. Cohen–Tannoudji, J. Dupont–Roc, and G. Grynberg, *Atom–Photon Interactions*, Wiley, 1992.

24. W.G. Harter, *Principles of Symmetry, Dynamics and Spectroscopy*, Wiley, 1993.

25. T.F. Gallagher, *Rydberg Atoms*, Cambridge University Press, 1994.

Papers:

Here we have listed some review articles and original papers related with the subjects of the present book (in chronological order).

1. D. Mendeleev, J. Russ. Phys.-Chem. Soc. **1**, 60(1869).

2. J.J. Balmer, Ann. Physik **25**, 80(1885).

3. J.R. Rydberg, Ann. Physik **58**, 674(1896).

4. P. Zeeman, Phil. Mag. **43**, 226(1896).

5. M. Planck, Ann. Physik **4**, 553(1901).

6. J.D. Van der Waals, Jr., Akad. Wetenschap. Amsterdam **11**, 79(1902).

7. G. Mie, Ann. Phys. **11**, 657(1903).

8. A. Einstein, Ann. Physik **17**, 132(1905).

9. F. Paschen, Ann. Physik **27**, 565(1908).

10. E. Rutherford, Phil. Mag. **21**, 669(1911).

11. N. Bohr, Phil. Mag. **26**, 1(1913).

12. H.G.J. Moseley, Phil. Mag. **26**, 1024(1913).

13. T. Lyman, Phys. Rev. **3**, 504(1914).

14. J. Frank and G. Hertz, Verhandl. Dent. Physik. Ges. **16**, 512(1914).

15. E. Schrödinger, Z. Physik **4**, 347(1921).

16. F.S. Brackett, Nature **109**, 209(1922).

17. W. Gerlach and O. Stern, Z. Physik **8**, 110(1922).

18. D.R. Hartree, Proc. Cambridge Phil. Soc. **21**, 625(1923).

19. A.H. Compton, Phys. Rev. **21**, 483(1923).

20. A.H. Pfund, J. Opt. Soc. Am. **9**, 193(1924).

21. H.A. Kramers, Nature **113**, 673(1924).

22. M. Born, Z. Physik **26**, 379(1924).

23. L. de Broglie, Phil. Mag. **47**, 446(1924).

24. H.N. Russell and F.A. Saunders, Astrophys. J. **61**, 38(1925).

25. W. Heisenberg, Z. Physik **33**, 879(1925).

26. G.E. Uhlenbeck and S.A. Goudsmit, Naturwiss **13**, 593(1925).

27. W. Pauli, Jr., Z. Physik **31**, 765(1925).

28. C.J. Davisson and L.H. Germer. Nature **119**, 558(1927).

29. M. Born and J.R. Oppenheimer, Ann. Physik **84**, 457(1927).

30. W. Heitler and F. London, Z. Physik **44**, 455(1927).

31. G.P. Thomson, Proc. Roy. Soc. (London) A **117**, 600(1928).

32. V. Fock, Z. Physik **49**, 339(1928).

33. P.M. Morse, Phys. Rev. **34**, 57(1929).

34. J.E. Lennard–Jones, Proc. Roy. Soc. (London) A **129**, 604(1930).

35. E. Hylleraas, Z. Physik **74**, 216(1932).

36. T.H. Koopmans, Physica **1**, 104(1933).

37. G. Racah, Phys. Rev. **61**, 186(1942).

38. C.C.J. Roothaan, Rev. Mod. Phys. **23**, 69(1951).

39. P.O. Löwdin, Rev. Mod. Phys. **32**, 328(1960).

40. U. Fano, Phys. Rev. **124**, 1866(1961).

41. O. Sinanoğlu, Phys. Rev. **122**, 491(1961).

42. E. Clementi and D.L. Raimondi, J. Chem. Phys. **38**, 2686(1963).

43. E.A. Hylleraas, Adv. Quantum Chem. **1**, 1(1964).

44. İ. Öksüz, Phys. Rev. A **13**, 1507(1976).

45. Ş. Erkoç and İ. Öksüz, Physica Scripta **20**, 658(1979).

46. Ş. Erkoç and İ. Öksüz, Chem. Phys. Letters **66**, 587(1979).

47. Ş. Erkoç, Phys. Stat. Sol. (b) **152**, 447(1989).

Index

www.ingramcontent.com/pod-product-compliance
Lightning Source LLC
Chambersburg PA
CBHW050635190326
41458CB00008B/2278